FORWARD-TIME POPULATION GENETICS SIMULATIONS

FORWARD-TIME POPULATION GENETICS SIMULATIONS

Methods, Implementation, and Applications

BO PENG
Department of Genetics, University of Texas,
M.D. Anderson Cancer Center

MAREK KIMMEL
Department of Statistics, Rice University

CHRISTOPHER I. AMOS
Department of Genetics, University of Texas,
M.D. Anderson Cancer Center

WILEY-BLACKWELL
A JOHN WILEY & SONS, INC., PUBLICATION

Published by John Wiley & Sons, Inc., Hoboken, New Jersey.
Published simultaneously in Canada.

Library of Congress Cataloging-in-Publication Data:

Peng, Bo, 1974-
Forward-time population genetics simulations : methods, implementation, and applications / Bo Peng, Marek Kimmel, Christopher I. Amos.
p. ; cm.
Includes bibliographical references and index.
ISBN 978-0-470-50348-5 (pbk.)
I. Kimmel, Marek, 1959- II. Amos, Christopher I. III. Title.
[DNLM: 1. Genetics, Population. 2. Biological Evolution. 3. Computer Simulation. 4. Models, Genetic. QU 450]
LC-classification not assigned
576.5'8–dc23

2011033593

Printed in the United States of America.

10 9 8 7 6 5 4 3 2 1

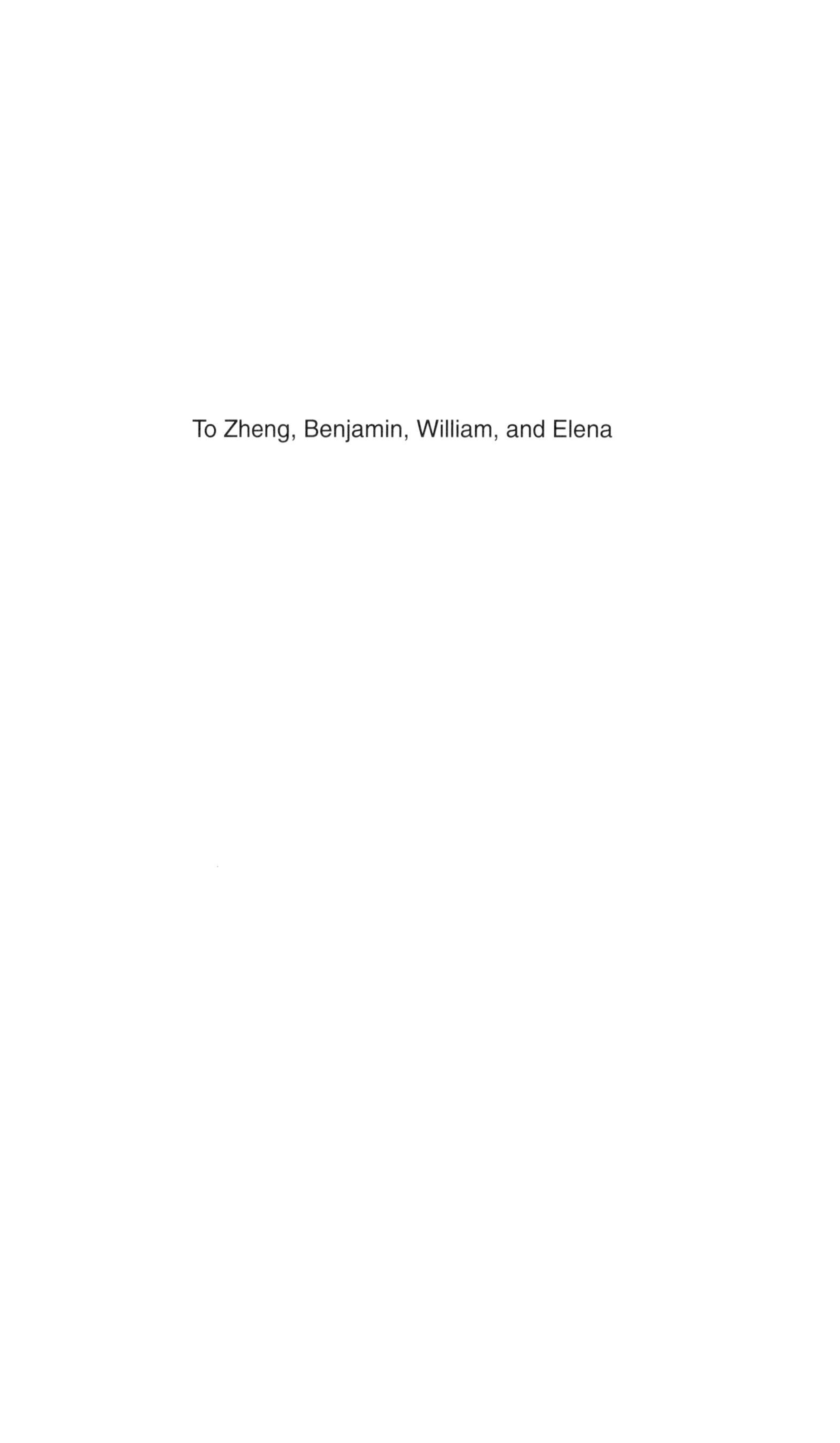

To Zheng, Benjamin, William, and Elena

CONTENTS

PREFACE

Forward-time population genetics simulation is simple in concept. Given a population with individuals of certain genotype, we evolve the population generation by generation, subject to various demographic and genetic forces such as population size change, mutation, selection, recombination, and migration. Population properties such as allele frequencies can be observed dynamically or be studied at the end of the simulation. Because this process mimics fundamental ways the human populations evolve, it is not surprising that such simulations have been used for decades and played an important role in the development and application of population and evolutionary genetics. However, due to the overwhelming demand for computing power in realistic population simulations, the applications of this simulation method have largely been limited to the development and demonstration of theoretical population genetics principles.

Recent years have witnessed a renewed attention to this old subject. Rapid developments in both methodology and software development have made forward-time population genetics simulation a promising tool to study complex evolutionary histories of different types of populations, with novel applications in the areas of population and evolutionary genetics, statistical genetics, genetic epidemiology, and even conservation biology. The revival of this simulation method can be largely contributed to two forces. The first is a strong need for highly flexible simulation method to simulate and study complex evolutionary histories. Although a large number of specialized methods are available, none of them is as flexible as forward-time simulations because forward-time simulations follow the direction at which populations evolve and can, at least in principle, simulate arbitrarily complex evolutionary scenarios. The second driven force is the continuous increase of the power of personal computers, which makes it possible to simulate millions of individuals for extended generations in a reasonable amount of time.

The fundamental advantage of forward-time simulations over other simulation methods is flexibility. Because this method is not restricted by any assumption, it can be used to simulate arbitrary complex evolutionary scenarios. However, despite the availability of a large number of simulation programs, very few of them can harness the full power of this simulation method. A typical forward-time simulation program is designed to simulate particular evolutionary processes for particular types of studies. Users are usually allowed to choose from a number of stocked genetic models and their parameters, but are not allowed to define their own evolutionary processes. For example, none of the existing programs can be used to study the evolution of a disease predisposing mutant, a process that is of great importance in statistical genetics and genetic epidemiology. Researchers who work on novel evolutionary models or new application areas without existing software are usually forced to write their own software.

The implementation of simuPOP was motivated by the studies of the evolutionary history of complex human diseases. Instead of a special-purpose program written for a few publications, this program was designed from ground up to be a general-purpose population genetics simulation program that can be used to simulate arbitrary evolutionary processes. Using a scripting language design, users of simuPOP could make use of many of its unique features, such as customized chromosome types, arbitrary nonrandom mating schemes, virtual subpopulations, information fields, and Python operators to construct and study almost arbitrarily complex evolutionary scenarios. This unique design makes simuPOP the best and in many aspects the only software packages for the simulation of complex evolutionary scenarios. Although some evolutionary scenarios could be simulated using other software packages, this book uses simuPOP to simulate all examples and lists source code of most examples so that users can learn how to implement various evolutionary scenarios and write their own simulations based on these examples. Note that although we describe most major features of simuPOP in the appendix of this book, this book is not a complete reference to simuPOP. Readers who would like to write complex scripts in simuPOP should refer to the simuPOP user's guide and reference manual for details.

Chapter 1 of this book gives an overview of important concepts and models that will be used in this book. Because of the mere number of concepts and models involved, they are introduced in a brief and often casual way. Interested readers should refer to standard textbooks on these subjects for more in-depth descriptions.

Chapter 2 simulates a number of standard population genetics models using a forward-time approach. The goal of these simulations is to demonstrate the impact of genetic factors such as mutation, selection, and recombination on standard Wright–Fisher models and how to use simuPOP to simulate them. Because detailed descriptions of these models are widely available in textbooks such as *Principles of Population Genetics* [1], we describe these models and their theoretical properties briefly, only as a way to motivate our simulations. Although simulations in this chapter are confirmatory in nature, they could be used to set up more complex evolutionary scenarios in which more than one genetic factor would be applied.

The rest of this book is devoted to applications of forward-time simulations in various research topics. Each chapter starts with a short description of the research topic and why forward-time simulations are used. The simulation processes are then described in detail. Because the primary focus of this book is on simulation techniques and not on particular research topics, we will present and discuss the results of these simulations briefly, leaving in-depth discussions to published papers on these topics. The simuPOP scripts that are used to perform all simulations are listed in the last sections of these chapters. Readers who are not interested in implementation details can safely skip these sections.

With continued increase of the power of personal computers and the availability of a powerful and flexible simulation engine, a wide range of interesting research topics could be attacked by forward-time population genetics simulations. We hope that this book can help researchers who are interested in such simulation design and implement their own simulations. We would welcome any comments and discussions and would appreciate the readers who would alert us to any errors they discover in this book.

BO PENG

Houston, Texas
2011

ACKNOWLEDGMENTS

The work covered in this book, especially the design and implementation of simuPOP, was done when the first author was a PhD student in the Department of Statistics at Rice University and a postdoctoral fellow in the Department of Epidemiology at the University of Texas, M. D. Anderson Cancer Center. The helpful and supportive comments of faculty and fellow students of the departments are hereby acknowledged.

A number of colleagues and students have helped in the development of simuPOP and in the writing of this book in various ways. Yaji Xu, a graduate research assistant, spent a lot of time on the documentation of simuPOP. His hard work during the summer of 2007 resulted in the first simuPOP release (0.8.0) that has a comprehensive online help system and a complete reference manual. Biao Li, a doctoral candidate in the Department of Bioengineering at Rice University, has helped in the development of allele frequency trajectory simulation functions and pedigree-related features of simuPOP and has written and executed some of the simulations for this book, especially the ones for Chapter 3. He also helped with the preparation of the bibliography and many figures of the book. Jianzhong Ma, PhD, read through the draft of this book and provided many useful suggestions. A high school student, Blake Kushwaha, helped proofread this book. They all deserve our sincere appreciation.

Numerous technical problems were encountered during the design and implementation of simuPOP and we relied on various online forums for help. We would especially like to thank the Python and SWIG (Simplified

Wrapper and Interface Generator, `http://www.swig.org`) user community, whose prompt replies to many e-mails were essential to the implementation of simuPOP.

User involvement was modest until early 2007, but has since then driven the development of simuPOP. Questions, bug reports, and feature requests from users have greatly enhanced the reliability and usability of this program and have led to the addition of many important features such as information fields and virtual subpopulations. One of the users, Tiago Antão, deserves a special thanks for his many bug reports and his contribution to the simuPOP online cookbook.

The development of a large software application such as simuPOP required a huge amount of time, many of which had to be drawn from time I should have spent with my wife Zheng Meng and our three children Benjamin, William and Elena. Their support during the past several years allowed me to pursue a career that I really enjoy, but has required many extra hours under the moonlight. I would like to dedicate this book to them.

Part of Bo Peng's research was supported by a training fellowship from the W.M. Keck Foundation to the Gulf Coast Consortia through the Keck Center for Computational and Structural Biology, and a Cancer Prevention Fellowship provided by the Jerry and Maury Rubenstein Foundation through the University of Texas, M.D. Anderson Cancer Center. Related research activities for all authors were partly supported by grant CA75432 from the National Cancer Institute, by grants ES09912 and R01CA133996-01 from the National Institutes of Health, and by grant 3T11F 01029 from Komitet Badań Naukowych (Polish Research Committee). Most of the simulations were performed using the Rice Terascale Cluster, funded by the National Science Foundation under grant EIA-0216467, by Intel, and by HP, and using the High Performance Cluster at the M.D. Anderson Cancer Center.

BO PENG

LIST OF EXAMPLES

CHAPTER 1

BASIC CONCEPTS AND MODELS

The simulation approaches that are described in this book involve knowledge from several disciplines. First, the genes and genomes are the targets of simulations, so some understanding of biology and genetics is needed. Then, the simulations involve the evolution of a collection of individuals over a long period of time, and we are concerned with the dynamics of the properties of the whole population rather than with a small number of individuals. This involves knowledge of population and evolutionary genetics. Finally, as the most important application area, we will simulate the evolution of human diseases and produce populations with affected individuals. Techniques from statistical genetics and genetic epidemiology will be used to locate genes that are responsible for the diseases.

This chapter reviews basic concepts and, more importantly, various mathematical models that will be used in this book, organized by disciplines. To target the most essential components, these concepts are often defined in a casual way that may not reflect their full biological or statistical complexity. For more in-depth descriptions and concrete examples, the reader should refer to standard textbooks on these topics [1–4]. Readers who are already familiar with one or more of the disciplines can skip relevant sections.

Forward-time Population Genetics Simulations: Methods, Implementation, and Applications,
Bo Peng, Marek Kimmel, and Christopher I. Amos.

1.1 BIOLOGICAL AND GENETIC CONCEPTS

1.1.1 Genome and Chromosomes

The genetic material of humans is called the human *genome*, which consists of 23 pairs of *chromosomes*. Humans are called *diploid* because we have two sets of chromosomes, one set of which was inherited from each parents. Some species, like bacteria, have only one set of chromosomes (called *haploid*), some plants have four (*tetraploid*), six (*hexaploid*), or more (*polyploid*) copies. Because this book concerns mostly human genomes and diseases, almost all examples simulate diploid populations.

Chromosomes are composed of *deoxyribonucleic acid* (DNA) molecules. DNA usually consists of two complementary chains twisted around each other to form a double helix. Each chain is a linear sequence of four nucleotides: adenine (A), guanine (G), cytosine (C), and thymine (T). Adenine pairs with thymine and cytosine pairs with thymine by means of hydrogen bonds. DNA plays two fundamental biological roles.

- DNA carries the instructions for making the components of a cell (mostly *proteins*). A single strand of DNA can act as a template for the enzymatic synthesis of a complementary strand of *messenger ribonucleic acid* (mRNA) through a process called *transcription*. The information encoded in mRNA is then translated to protein during a complex *translation* process that takes place in the cell's ribosomes. If anything wrong happens in the DNA that interrupts or changes this process, the body may not get the right amount of certain protein and show symptoms of a disease.
- Information encoded in DNA can be passed to daughter cells when a cell divides. During *meiosis* (the process during which gametes are produced as the result of DNA replication and two rounds of cell divisions of germline cells), DNA is replicated and used to form daughter cells. For humans, the inheritance pattern follows the *Mendelian Law*, that is, gametes contain one of the two sets of parental chromosomes, and offspring are formed by two parental gametes.

The lengths of double-stranded DNA molecules are described in units of base pairs, and for longer molecules in kilobase pairs (kb) or megabase pairs (Mb). Human chromosomes vary greatly in length and are numbered roughly in the order of their lengths. The longest chromosome (chromosome 1) is of about 263 Mb, and the shortest one (chromosome

21) is about 50 Mb. The overall size of the human autosomes is around 3093 Mb.

1.1.2 Genes, Markers, Loci, and Alleles

A *gene* is a specific region of DNA that codes for a single protein or enzyme. It is composed of a set of three adjacent nucleotides (a *codon*). These 64 different types of *codons* correspond to 20 kinds of *amino acids* that are the building blocks of proteins. A gene can be long (some genes span several Mb) and have complex structures. The most important aspects for genetic simulations are the location and variations occurring within or near a gene.

Genetic markers are DNA sequences that can be identified by a variety of biological techniques. Genetic markers are useful if they are *polymorphic*, meaning there is population variation at the marker. A marker may be short, such as a single base pair change (*single nucleotide polymorphism*), (*SNP*), or long, like *microsatellites*, which are short regions of tandemly repeating DNA sequence. Genes and markers are related, but are different concepts: a physical gene can have multiple markers, and a marker does not have to be inside a gene. Genes perform biological functions and can contribute to diseases, and they can be *homomorphic* (having no population variation). Markers do not have to be functional, but need to have a known location and are usually required to be *polymorphic*.

The position of a gene or marker on a chromosome is known as its *locus* (the plural form is *loci*). Variants of the DNA sequence at this locus among individuals are called *alleles*. If a marker (e.g., a SNP marker) has two alleles, it is called *diallelic*. If an individual carries the same alleles on both DNA strands at a locus, he is said to be *homozygous* at this locus. Otherwise, he is *heterozygous* at this locus. Generally speaking, at each locus there is a *wild-type* allele that is thought to be result in the wild or normal phenotype. In this book, all alleles are coded as numbers. The wild-type allele is often coded as allele 0, and others as allele 1, 2, 3, ...

The DNA sequence of interest is the *genotype* of an individual. The physical expression of a genotype is called the *phenotype*. For example, some genes control the color of our eyes. These genes are the genotype of the phenotype eye color. Note that the underlying relationship between DNA sequence and phenotype is more complex than such one-to-one or many-to-one correspondences, but for all the purposes of this book, we assume that one or several genes cause a single phenotype, which can often be observed as a *quantitative trait* such as blood pressure or the *affection status* of a disease.

1.1.3 Recombination and Linkage

Genetic *recombination*, also called *crossing over*, refers to genetic events that can occur during the formation of sperm and egg cells. During the early stages of cell division in meiosis, two chromosomes of a homologous pair may exchange segments, producing genetic variations in germ cells. For example, if one homologous chromosome has a *haplotype* (genetic sequence on the same chromosome) AB, and another homologous chromosome has a haplotype ab, one of the gamete cells, because of recombination, may have a chromosome with genotype Ab. Such gametes are called *recombinants*. The proportion of recombinants is called the *recombination rate* between these two loci, which is $\frac{1}{2}$ if two loci are on two different chromosomes, and thus segregate independently. In addition to the *independent assortment* of chromosomes, which leads to 2^{23} different types of gametes due to random choices of chromosomes, recombination leads to more variations among gametes, and therefore variations among offspring of the same parents.

The *genetic distance* (also called *map distance*) between two loci is defined as the average number of crossovers between the loci per meiosis. The unit of genetic distance is the centiMorgan (cM). Two loci are 1 cM apart if on average there is one crossover occurring between these two loci on a single strand for every 100 meiosis. The distribution of recombination events varies between the sexes: females have on average 1.65-fold more recombination events when they make eggs than males do when they make sperms. Even on a single chromosome, recombination rate is uneven, and there exists *recombination hotspots* with peak recombination rate hundreds or thousands times that of the surrounding regions [5].

Because of the uneven recombination rates across chromosomes, the map distance does not have to reflect the true *physical distance* between them, which is measured in base pairs. As a genome-wide average for loci, 1 cM roughly corresponds to one million base pairs (1cM/Mb). This provides a rough estimate of the distance between two markers if their genetic distance is known and small. For longer map distances, because the occurrence of multiple crossovers between two loci can no longer be ignored, various *map functions* should be used. The *recombination fraction* measures the probability that two alleles will be observed to show a crossover event and there is probability that an odd number of crossovers will occur between the two loci, since an even number of crossovers would not lead to an observable reshuffling of genetic material at the two loci. The simple $x = \theta$ formula where x is physical distance and θ is map distance is known as *Morgan's map function*. This function is accurate only for very short distances in which the probability of an even number of crossovers is very low. Other frequently used map functions include the *Haldane map function* $x = -\frac{1}{2}\ln(1-2\theta)$ and *Kosambi map function* $x = \frac{1}{4}\ln\frac{1+2\theta}{1-2\theta}$.

Despite the random occurrence of recombination events, the general rule holds that if two loci are physically close to each other, they tend to cosegregate during meiosis because of the low probability of crossing over between them. This property enables the use of a marker to determine the inheritance pattern of and potentially the location of a disease-predisposing gene that has previously been localized. The effectiveness of this method depends on two related concepts: *linkage* and *linkage disequilibrium* (LD). Two markers are *linked* if the recombination fraction between them is less than $\frac{1}{2}$, and are *unlinked* if they segregate independently (the recombination rate equals $\frac{1}{2}$). Linkage is a measure of how closely two markers, or a marker and a disease gene, are located on a chromosome. This concept is related to, but different from, how the alleles at a marker are associated with the disease alleles in the population (LD between the marker and the disease gene). We will introduce the exact definition of LD later.

1.1.4 Sex Chromosomes

The first 22 pairs of human chromosomes are identical between male and female, they are called *autosomes*. The members of the last pair are known as the *sex chromosomes* because they differ between the sexes. Females have two copies of the *X chromosome*, and males have one *X* chromosome and a smaller *Y chromosome*. Because *X* and *Y* chromosomes have different lengths (and genetic material), *X* and *Y* chromosomes cannot recombine over their entire lengths as autosomes do. In fact, only a small portion (<10%) of the *X and Y* chromosomes (*pseudoautosomal regions*) can recombine during spermatogenesis. The *X chromosome* in females behaves like an autosome (and can recombine along the entire length with another copy of *X* chromosome).

1.1.5 Mutation and Mutation Models

Any change of genetic material is called *a mutation*. It is the source of genetic variation in the human population. Mutation can be caused by the substitution of a single base in the genome, small insertions and deletions of a few bases, expansions or contractions in the number of tandemly repeated DNA motifs, insertions, deletions, duplications, and inversions of long segments of DNA, translocation of chromosomal segments, and even changes in chromosomal number. The probability that a mutation happens at a locus is called the *mutation rate* at this locus. Mutation rates can differ from locus to locus. If we consider a single nucleotide as a locus, the mutation rate is below 10^{-8} per locus, per generation. Microsatellite markers have a higher mutation rate, which is up to 10^{-4}. Because mutations may have different effects on different alleles, there can be several mutation rates influencing variability at a single locus.

Mutations will change the genotype of an individual, but not necessarily the phenotype. For example, because multiple codons can code the same amino acid, a mutation may change a codon without altering the encoded amino acid. Such a mutation is called a *silent* or *synonymous* mutation. A *nonsynonymous mutation* changes the codon in such a way that gene product is changed. Mutation can happen in *somatic* (body) or *germline* cells, but *somatic mutations* will not change the evolutionary process. These mutated cells may propagate and cause diseases such as cancer, but as long as the *germline* cells are unaffected, the mutation will not be passed to gametes.

Despite all the complexities behind mutation events, mutation in this book is simply a change from one allele to another allele at a marker locus. Because we do not restrict the types of markers, this concept can encompass most mutation events, except for chromosome number changes, which are usually so deleterious that they lead to offspring that are not reproductive in humans. For example, mutation at a SNP marker can change the nucleotide at this marker to another one, mutation at a microsatellite marker can produce a new allele with another number of short tandem repeats, a mutation that knocks out a whole gene can be considered as creating a new null allele.

If we categorize alleles into wild-type alleles and disease alleles, a mutation from a wild-type allele to a disease allele is called *forward mutation*, and a mutation from a disease allele to a wild-type allele is called *back mutation*. Because there are far more ways to damage the proper function of a gene than to restore it, back mutations are frequently assumed to happen at a much lower rate than forward mutations. When there are a large number of disease alleles, we can assume an *infinite allele mutation model* so that each mutation event will generate a different mutant. There is another frequently used model called *infinite site mutation model*, which assumes that each mutation happens at a different site of a long sequence. This model is equivalent to the infinite allele mutation model if we consider the whole sequence as a marker.

If the number of allele states is limited, a much more realistic model is the k-allele model. Under a k-allele model, a locus can have at most k alleles. When a mutation happens, an allele has probability $\frac{1}{k-1}$ to become any other allele. If we treat allele 0 as wild-type allele and all others as disease

TABLE 1.1 Kimura's Two Parameter Mutation Model

Nucleotide	A	G	C	T
A	$1-\alpha-2\beta$	α	β	β
G	α	$1-\alpha-2\beta$	β	β
C	β	β	$1-\alpha-2\beta$	α
T	β	β	α	$1-\alpha-2\beta$

alleles, the probability of mutating from a wild-type allele to disease allele is 1 and the probability of mutating from a disease allele to a wild-type allele is $\frac{1}{k-1}$. When k is large, the back mutation rate is very small, so we can approximate the infinite allele model with a k-allele model. At the other extreme, a *two-allele* model can be used to model the mutation between the two alleles of a SNP marker.

If the assumption that an allele can mutate to any other allele at equal probability does not hold, more complicated models can be applied. For example, due to biological reasons, the rate of *transitions* between purines ($A \leftrightarrow G$) or pyrimidines ($C \leftrightarrow T$) may be different from the rate of *transversions* between a purine and a pyrimidine ($A, G \leftrightarrow C, T$). *Kimura's two parameter model* models these differences using two parameters with mutation rates displayed in Table 1.1.

Microsatellite markers mutate in a different manner. During the mutation, the number of tandem repeats will change by a small number as the result of expansions or contractions. The simplest model is the *symmetric stepwise mutation model* where allele a mutates to allele $a+1$ or $a-1$ with equal probability. Extensions to this model exist. For example, a *generalized stepwise mutation model* mutates allele a to $a+n$ or $a-n$ where n is drawn from a random distribution. A *geometric generalized stepwise mutation model* assumes a geometric distribution for n.

1.2 POPULATION AND EVOLUTIONARY GENETICS

The genetic composition of our human population is the result of a long and complex evolutionary process. Studying this evolutionary process can provide valuable information about our current population. For example, without any ancestry information, our human population would present to us as 6 billion independent individuals and there would be no way to explain why some people are genetically more similar to each other than others.

An enduring goal of population and evolutionary genetics is to understand the forces that govern how populations and species evolve. There are many such forces such as *demographic* changes (notably the changes of population size) and genetic factors such as recombination, mutation, migration, genetic drift, and selection. All these forces can leave signals on the current population and we are often interested in inferring the existence, time, intensity, and duration of these forces. Among these forces, researchers are especially interested in elucidating the relative contributions of genetic drift and natural selection to extant patterns of genetic variation [6, 7]. More specifically, the neutral theory posits that most polymorphisms are either neutral or slightly deleterious and changes in allele frequency are primarily governed by the stochastic effects of genetic drift in populations of finite size. An alternative view is that a significant proportion of variation does affect the ability of an organism to survive and reproduce and will therefore be subject to natural selection.

1.2.1 Population Variation and Mutation

Population variation refers to the phenotypic and genotypic variations among humans. We all differ from each other phenotypically, in features such as skin color and height. Many of these differences have a genetic basis and tend to "run in families." The ultimate source of genetic variation is mutation. We often assume that initially all individuals in a population have the same genotype at a locus, which is called the *wild-type* allele, until a *mutant* is introduced to the population as a result of mutation. The new mutants will then spread in the population according to Mendels law of independent assortment. The mutant alleles usually become extinct, but some of them can reach higher allele frequency over time. The forces that can help maintain a mutant in a population or reach higher allele frequency include *balancing selections* (such as *heterozygous advantage*), strong genetic drift caused by small population size or bottlenecks, positive selection, or being linked to a gene under positive selection (the *hitchhiking* effect). These forces are important in the simulation of human genetic diseases.

1.2.2 The Wright–Fisher Model and Random Mating

The simplest case of a *Wright–Fisher model* is based on a haploid population of fixed size N. Assume that there is no selection or mutation, and the genes in the offspring generation are derived by sampling with replacement from the parental generation. Assume that there are $X_t = i$ copies of allele a in generation t, the distribution of X_{t+1} follows a binomial distribution.

Namely,

$$\Pr(X_{t+1} = j) = \binom{N}{j}\left(\frac{i}{N}\right)^{j}\left(1 - \frac{i}{N}\right)^{N-j}, \quad i, j = 0, 1, ..., N. \quad (1.1)$$

The basic features of Wright–Fisher models are *random mating* and *nonoverlapping generations* and all other assumptions can be relaxed. If the population is diploid, there will be $2N$ genes, so all N in Equation (1.1) need to be replaced by $2N$. Population size does not have to remain constant from generation to generation. If two sexes are considered, mating between males and females is assumed to occur randomly from their respective groups. If selection is considered, individuals will be chosen randomly with a probability that is proportional to its fitness value.

Random mating is an idealized mating scheme that implies that an individual has equal probability to mate with anyone else in the population. This is of course far from reality. Human populations have complex structures and many factors (demographic or social) that prevent the free flow of genes within and between populations. Although most simulations that will be discussed in this book use random mating schemes to simulate the Wright–Fisher models, nonrandom mating schemes will be used occasionally to simulate more complex evolutionary processes.

1.2.3 The Hardy–Weinberg Equilibrium

Under the assumptions of random mating with sexual reproduction, nonoverlapping generations, and no migration, mutation, and natural selection, the genotype and allele frequencies of a diallelic locus in an infinitely sized diploid population remain constant from generation to generation. As a result, Hardy–Weinberg equilibrium occurs and the distribution of genotypes in the population can be described by the following simple mathematical relation:

$$AA : p^2, \quad Aa : 2pq, \quad aa : q^2, \quad (1.2)$$

where p^2, $2pq$, and q^2 are the frequencies of the genotypes AA, Aa, and aa in zygotes of any generation, p and q are the frequencies of A and a in gametes of the previous generation, and $p + q = 1$. Hardy–Weinberg equilibrium indicates a simple conversion between allele and genotype frequencies and is widely used in theoretical models in population genetics. The mathematical derivation of Equation 1.2 can be found in many population genetics textbooks.

1.2.4 Genetic Drift and Effective Population Size

Genetic drift is the random change in allele and haplotype frequencies in populations of finite size, as a result of random sampling of gametes from generation to generation. In a large population, on an average, only a small change in the allele frequency will occur. However, when the population size is small, genetic drift can lead to rapid and random change of allele frequencies, and can result in fixation or the loss of an allele. Genetic drift is an important factor when designing a forward-time simulation. For example, even if a disease allele has the same initial frequencies in two populations at the beginning of a set of replicate simulations, the allele frequencies might differ greatly at the end, making direct comparisons between the final populations difficult.

According to the assumptions used to develop the Wright–Fisher law, all individuals are involved in the transmission of genetic materials from parental to the offspring generation. This is not the case for human populations, partly because not all humans are in reproductive ages and also because not all individuals reproduce. That is to say, the *breeding population sizes* of human populations are smaller than their *census population sizes*. When we perform a forward-time simulation, there is no need to simulate a large population with only a small percent of genetically active individuals. It is therefore more practical to simulate a random mating population in the size of an idealized Wright–Fisher population that contains the same amount of genetic drift observed in the actual population under consideration. The size of such an idealized population is called the *effective population size* of a given population, which plays an essential role in population genetics studies [8]. There are many methods to estimate the effective population size of the human populations, using different properties of the Wright–Fisher model. The estimated effective population size of the human population is on the order of 10^4, far less than its census size 6×10^9.

On the other hand, not all features of a natural population can be captured by a Wright–Fisher population with an appropriate effective population size. Because forward-time simulations are capable of simulating nonrandom mating schemes in natural populations, they can be used to study complex evolutionary processes with unknown or varying effective population sizes.

1.2.5 Natural Selection

Natural selection is a process that favors or induces survival and perpetuation of one kind of organism over others. Selection can be *positive* (or *advantageous*) or *negative* (or *purifying*) and has a profound impact on the evolution of the human population. In addition, selection can be balancing in which the genotypes have a mixture of positive and negative selection pressures so that there is no net effect of selection on the individual alleles. The central concept of natural selection is *fitness*, namely, the ability of organisms to survive and reproduce in the environment in which they find themselves [9]. Although the concept of fitness can be quite involved, this book uses the simplified version of two fitness concepts: *absolute fitness* and *relative fitness*.

The *absolute fitness* is a statistic that is used to summarize the total fitness, namely, viability, mating success, fecundity, and so on of individuals. In terms of simulation, the absolute fitness used in this book is the probability of surviving of offspring. For example, if a mating event produces 10 offspring, each with an absolute fitness of 0.8, the number of surviving offspring follows a binomial distribution with parameters 10 and 0.8. Because absolute fitness describes the absolute viability of individuals, the survival probability of one individual is independent of the probabilities of others.

Although absolute fitness is easy to think about, a different statistic called *relative fitness*, is almost always used. The relative fitness, of an individual equals its absolute fitness normalized in some way and can be understood as the relative ability of an individual to pass his or her genotype on to a future generation comparing a genotype with a referent genotype. We model relative fitness as the relative probability to mate (which we will call selection against parents) due to its simplicity in implementation.

If we assume that selection acts only on a single DSL, natural selection can be modeled by the relative fitness of genotypes AA, Aa, and aa. Assume that genotypes AA, Aa, and aa have population frequency P_{AA}, P_{Aa}, and P_{aa}, and relative fitness w_{AA}, w_{Aa}, and w_{aa}, respectively. Assume that n offspring are produced and w_{ij} is the survival rate of offspring with genotype ij. $nP_{ij}w_{ij}$ offspring with genotype ij will survive and lead to

$$P'_{ij} = \frac{P_{ij}w_{ij}}{P_{AA}w_{AA} + P_{Aa}w_{Aa} + P_{aa}w_{aa}}, \tag{1.3}$$

where P'_{ij} is the genotype frequency of genotype ij in the offspring generation . Now, using the "ability-to-mate" approach with sexless random mating, the proportion of genotype ij in the offspring generation is the number of ij individuals times its probability to be chosen:

$$P'_{ij} = n_{ij}\frac{w_{ij}}{\sum_{n=1}^{N} w_n}, \tag{1.4}$$

where $N = n_{AA} + n_{Aa} + n_{aa}$ is the size of the parental generation. P_{ij} in Equation 1.4 equals to that is Equation 1.3 because

$$n_{ij}\frac{w_{ij}}{\sum_{n=1}^{N} w_n} = n_{ij}\frac{w_{ij}}{n_{AA}w_{AA} + n_{Aa}w_{Aa} + n_{aa}w_{aa}}$$
$$= \frac{P_{ij}w_{ij}}{P_{AA}w_{AA} + P_{Aa}w_{Aa} + P_{aa}w_{aa}}.$$

This implies that at least in the case of a sexless Wright–Fisher model, natural selection that selects against parents using relative fitness values is equivalent to natural selection that selects against offspring using their absolute fitness values.

We call a fitness (selection) model *additive* if the fitness relationships for genotypes AA, Aa, and aa are 1, $1 - s/2$, and $1 - s$, *recessive* for the case of 1, 1, and $1 - s$, and *dominant* for the case of 1, $1 - s$, and $1 - s$. These models can be generalized using a *general dominance* model with fitness relationships 1, $1 - hs$, and $1 - s$ for genotypes AA, Aa, and aa. Other models include *heterozygous advantage* or *heterozygous disadvantage*, meaning the fitness of heterozygotes (Aa) is higher or lower than the fitness of AA and aa, respectively. Heterozygous advantage tends to maintain multiple alleles in a population, so it belongs to a broader definition of *balancing selection* defined as a type of selective regime in which multiple alleles at a locus are maintained in a population. On the contrary, *purifying selection* removes deleterious alleles from a population, and *positive selection* favors certain alleles. Purifying selection tends to keep only the wild-type allele in the population, while positive selection will lead to the removal of the wild-type allele in favor of the mutant allele. Note that neutral alleles may be carried along because they are in close vicinity of another locus under positive selection. This phenomenon is called *genetic hitchhiking*, which can cause *selective sweeps* in which a selectively advantageous allele increases in frequency and results in changes of the frequency of alleles that are in linkage disequilibrium with it.

Selection pressure does not have to be constant over the lifetime of an individual or over the evolutionary history of the human population. For example, some diseases like Alzheimer show decreased fitness only at the later part of human lives. Because such diseases may not affect the fitness before the mating age, they may show no overall selective disadvantage. Another example is that a disease may be advantageous at first, but at a cost of deteriorated fitness later. Such a model is called *antagonistic pleiotropy*. The selection pressure on a disease allele may also change because of environmental and social changes.

TABLE 1.2 An Example of Frequency-Dependent Selection

Fitness	BB	Bb	bb
AA	1	1	1
Aa	0.999	0.99	0.9
aa	0.998	0.98	0.8

Selection can also act on several loci, the fitness of the individual is then determined by the genotypes at all loci. Theoretical studies sometimes treat these loci separately and apply single-locus fitness model to each locus independently [10]. However, because the overall fitness of an individual is the combined effect of multiple loci, relative fitness at individual locus cannot be used separately in simulations.

Simple multilocus selection models combine single-locus fitness values to produce an overall fitness value. The overall fitness of an individual with fitness g_i, $i = 1, ..., L$, at each locus is $g = \prod_{i=1}^{L} g_i$ for a *multiplicative multilocus selection model* and $g = 1 - \sum_{i=1}^{L} (1 - g_i)$ for an *additive* one. In such a model, the *marginal fitness* at a locus, meaning the population average of the fitness of individuals having certain genotype at a locus, is close to the single-locus model at this locus. Therefore, from a population point of view, these models lead to roughly independent evolution of individual locus.

The disease loci can interact with each other, and with environmental factors in many ways. It is difficult to analyze these models because marginal fitness can change because of the changes of environment factors or allele frequency at other loci. The change of fitness as a result of different genotype frequencies within a population is called *frequency-dependent selection*. This type of selection is observed in many natural environments and may be explained by interaction between disease loci. Such an example is given in Table 1.2 where locus B acts as a modifier to locus A. Locus B does not cause the disease by itself, but decreases the fitness of individuals with allele a, from slightly (Aa or aa with heterozygote Bb) to highly deleterious (with homozygote bb). Therefore, the total disease allele frequency changes with respect to the frequency of allele b, and we may no longer have constant selection pressure over locus A.

1.2.6 Linkage Equilibrium

Linkage disequilibrium is the nonrandom association of alleles between two or more loci. Although it is commonly used to measure correlations between linked loci, linkage disequilibrium can be detected between unlinked loci because non-random associations can arise due to many reasons, including population structure, admixture, and epistatic selection. It is sometimes desired to differentiate linkage disequilibrium caused by linked loci from associations between arbitrary loci.

In a diploid population, two alleles A and a are segregating at locus A, and alleles B and b are segregating at a second locus B. There are then four possible gametes ab, aB, Ab, and AB. Let their frequencies in the gametic pool be p_{ab} and so on. If loci A and B are *unlinked*, meaning that alleles A and a segregate independent of B and b, and there is no spurious association between these two loci, the haplotype frequency p_{ab} should equal to $p_a p_b$. This case is called *linkage equilibrium*. If loci A and B are linked, alleles at these two loci tend to cosegregate. For example, if allele a is associated with allele b, the genotype frequency p_{ab} may no longer equal to the product of allele frequencies. To measure the level of such cosegregation, we define *linkage disequilibrium* as

$$D = p_{ab} - p_a p_b, \tag{1.5}$$

when both loci are *diallelic*, meaning that only two alleles exist at each locus. If additional alleles are present at either locus, then this definition is extended to encompass haplotypic and allelic frequencies, but multiple values of the disequilibrium coefficient will be needed to characterize the loci.

A drawback of using D to measure association between loci is that the range of D depends on the frequency of these gametes. To fix this problem and have a measure that ranges between 0 and 1, another measure of linkage disequilibrium

$$D' = \frac{D}{D_{\max}} \tag{1.6}$$

is commonly used, where $D_{\max}$ is the theoretical maximum for the observed allele frequencies.

$$D_{\max} = \begin{cases} \min\left(p_A(1 - p_B), (1 - p_A)\, p_B\right), & \text{if } D > 0, \\ \min\left(p_A p_B, (1 - p_A)(1 - p_B)\right), & \text{otherwise.} \end{cases}$$

Other frequently used measures of linkage disequilibrium include

$$r^2 = \frac{D^2}{p_A(1 - p_A)\, p_B(1 - p_B)}. \tag{1.7}$$

Because the sign of D depends on the choice of alleles, the absolute values of D are used in this book unless alleles to be used are specified at both loci.

When there are more than two alleles at one or more loci, the average of pairwise LD values are used. That is to say,

$$D = \sum_{i,j} p_{a_i} p_{b_j} \left| p_{a_i b_j} - p_{a_i} p_{b_j} \right|, \tag{1.8}$$

$$D' = \sum_{i,j} p_{a_i} p_{b_j} \left| \frac{p_{a_i b_j} - p_{a_i} p_{b_j}}{D_{\max}(i, j)} \right|, \tag{1.9}$$

$$r^2 = \sum_{i,j} p_{a_i} p_{b_j} \left| \frac{\left(p_{a_i b_j} - p_{a_i} p_{b_j}\right)^2}{p_{a_i}\left(1 - p_{a_i}\right) p_{b_j}\left(1 - p_{b_j}\right)} \right|, \tag{1.10}$$

where $\sum_i \sum_j$ iterate through all alleles a_i at locus A and b_j at locus B. $D_{\max}(i, j)$ is defined analogous to $D_{\max}$. Note that other definitions (e.g., use haplotype frequency $p_{a_i b_j}$ instead of allele frequencies $p_{a_i} p_{b_j}$) have been used for multiallelic LD measures and Equations 1.8–1.10 are the ones that are implemented in this book. Because the sign of D depends on the choice of reference alleles at loci A and B, we use Equations 1.8 and 1.9 to obtain positive LD values unless reference alleles are specified at each locus.

A disease mutant, when it is first introduced to the population, may have high LD with its surrounding markers, particularly if the disease mutant (a) occurs on a haplotype that has the minor allele (b) at the second locus. Linkage disequilibrium will deteriorate when recombination happens between loci A and B. It is proved that with recombination rate r, assuming a large population, mating at random,

$$D_t = (1 - r) D_{t-1} = \cdots = (1 - r)^t D_0, \tag{1.11}$$

so linkage disequilibrium will decay to zero eventually in a population of infinite size. The rate of such decay depends on recombination rate, namely, the genetic distance, between two loci. If the population is not large or mating at random, the decay will be more gradual, but will still occur if mating does not strictly occur according to the mutant allele type.

1.2.7 Population Structure and Migration

Until now, we have assumed that mating in a population is random. That is to say, any individual will have the same probability to mate with any individual of the opposite sex. This does not hold in the presence of *population structure*, when a population is divided into several *subpopulations*. Mating can be assumed to be random within subpopulations, but is not allowed between subpopulations. Because of the stochastic nature of gene flow, subpopulations may have distinct genetic features even if they evolve from the same founder population.

Migration can exchange individuals, therefore genetic materials, between subpopulations. *Migration rate* usually refers to the percentage of individuals that migrate to other subpopulations. Two migration models that have been studied extensively in theoretical population genetics are the *island model* and the *stepping stone model* [11]. Assuming n subpopulations numbered from 1 to n, an island model allows migration from any subpopulation to any other subpopulation. The migration rate matrix can be written as

$$r_{ij} = \begin{pmatrix} 1-r & \frac{r}{n-1} & \cdots & \cdots & \frac{r}{n-1} \\ \frac{r}{n-1} & 1-r & \cdots & \cdots & \frac{r}{n-1} \\ \cdots & \cdots & \cdots & \cdots & \cdots \\ \frac{r}{n-1} & \cdots & \cdots & 1-r & \frac{r}{n-1} \\ \frac{r}{n-1} & \cdots & \cdots & \frac{r}{n-1} & 1-r \end{pmatrix},$$

where r_{ij} is the percentage of individual that will migrate from subpopulation i to j.

A *stepping stone model* allows migration only between adjacent subpopulations, that is, from subpopulation i to $i-1$ and $i+1$. This leads to a migration rate matrix

$$r_{ij} = \begin{pmatrix} 1-r & r & \cdots & \cdots & \cdots \\ r/2 & 1-r & r/2 & \cdots & \cdots \\ \cdots & \cdots & \cdots & \cdots & \cdots \\ \cdots & \cdots & r/2 & 1-r & r/2 \\ \cdots & \cdots & \cdots & r & 1-r \end{pmatrix}.$$

Following the same idea, two-dimensional stepping stone models can be used to model the migration to adjacent cities of human populations.

1.2.8 Demographic History of Human Populations

Demographic history of a population has strong impact on the genetic composition of the current generation. For example, families in a population that has gone through rapid population expansion tend to be large because parents in such a population tend to have many offspring. A population may have reduced genetic variation and skewed allele frequency spectrum of polymorphisms if it experienced a *population bottleneck*, which refers to a reduction in population size that increases the effects of genetic drift and reduces genetic variation. The effect of a bottleneck on patterns of genetic variation depends on how severe the decrease in population size is and the duration of the bottleneck. Other important demographic factors include migration and *population admixture*. The latter refers to populations origin from two or more genetically separated populations.

The human populations have gone through a complex migration and settlement process. Modern human morphology was first found in Africa about 130,000 years ago, and only substantially later in other parts of the world. This is believed to be the result of a "Out of Africa" process, which consists of potentially several rounds of migration events. Some evidence suggest that there was a brief adventure around 90,000 years (4000–4500 generations) ago [12]. Then, there was an early "southern route" dispersal from the Horn of Africa around 55,000–85,000 years ago (around 3000 generations), along the tropical coast of Indian Ocean to Southeast Asia and Australasia. The real "out of Africa" migration to Eurasia happened around 45,000–55,000 years (2000–2500 generations) ago. During the migration, the founder population is split into subpopulations and migrate to different geographic regions, driving other humans to complete genetic extinction. This process can be modeled by a sequential colonization process in which a subpopulation migrates only to its adjacent subpopulations [13]. After subpopulations settled down, the population size usually expanded quickly. An exponential population growth model can largely be used, although some researchers prefer a logit model. Despite all the complexity and uncertainties in the demographic models of real human populations, examples in this book usually use a constant population size model or an exponential population expansion model.

1.2.9 Coalescent and Backward-Time Simulations

The coalescent is a stochastic model of gene genealogies that has become the central theoretical tool in population genetics for understanding, interpreting, and simulating genetic variation [14]. It provides not only a theoretical framework from which all sorts of statistical inferences can be made from observed genotype data but also a simulation method that can be used to simulate genetic samples efficiently.

A simple coalescent process starts with n alleles in a population of size $2N$. The process is followed backward in time until a common ancestor of two allele is found. Because the probability that alleles genes find a common ancestor in the parental generation is $\frac{1}{2N}$ (the first gene can choose its parent freely, the second gene must choose the same parent as the first gene) and $\left(1 - \frac{1}{2N}\right)^{j-1} \frac{1}{2N}$ in the j generations back in time if no common ancestor is found in the last 1, 2,. . . , $j - 1$ generations, the coalescence time T_2 for two alleles to find a common ancestor follows a geometric distribution with parameter $\frac{1}{2N}$:

$$\Pr(T_2 = j) = \frac{1}{2N}\left(1 - \frac{1}{2N}\right)^{j-1},$$

which can be approximated by an exponential distribution. The mean of T_2 is therefore $2N$ generations. Similarly, the probability that two genes out of the k genes find a common ancestor $T_k = j$ has approximately a geometric distribution with parameter $\binom{k}{2}/(2N)$. These two alleles will coalesce as a parental allele and the process continues with $n - 1$ alleles. The process stops when a single allele is found, which is referred to as the *most recent common ancestor* (MRCA) of all alleles. By simulating the coalescence of k alleles, a (random) gene genealogy of these k alleles can be constructed (Figure 1.1).

The coalescent process can be used to model genetic variation of samples because the parents of all alleles belong to this coalescent tree, and the genotypes of all genes in this tree can be determined regardless of the genotype information of the rest of the populations. The patterns of genetic variation are shaped by two stochastic events: (i) the history of coalescent

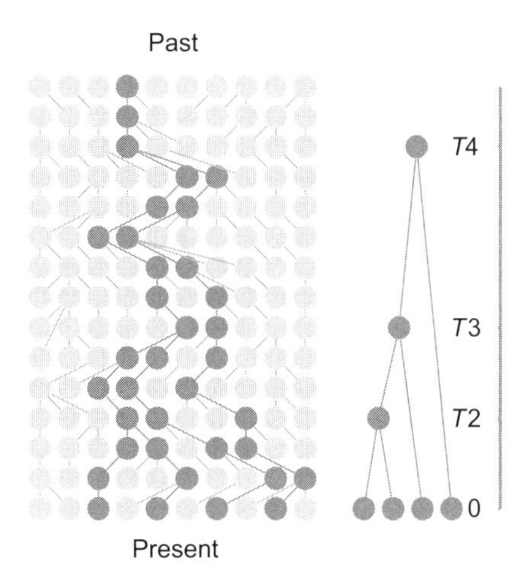

FIGURE 1.1 Illustration of the coalescent process.

events, which can be modeled as exponential random variables; and (ii) the history of mutational events, which can be modeled as Poisson random variables. Under the assumption of *neutrality* (no selection), mutations are uniformly distributed along the branches of a genealogy, and therefore the number of mutations occurring on a branch is proportional to its length (i.e., time to coalescence). The neutrality assumption plays a crucial role in this process because it allows these two stochastic events to be applied independently. That is to say, a coalescent tree can be constructed before mutations can be applied to the branches of this tree using arbitrary mutation models. This cannot be done with selection because the length of a branch would then depend on the number of mutations on this branch.

The coalescent process provides a model that can be used to build test statistics and make statistical inferences. For example, the properties of a sample generated under a (neutral) coalescent process can be used to test the existence of natural selection. Given a collection of genotype data, the coalescent process can be used to describe the data in terms of possible previous events. Because different data sets tend to have different probabilities of being generated for different parameters, they can be used to infer these parameters. The most well-known example might be the dating of Mitochondrial Eve, which is estimated from the *coalescent times* to the MRCA of worldwide samples of human mitochondrial DNA (mtDNA). mtDNA is a single, nonrecombining molecule that is unlikely to be under selection. Because mtDNA is exclusively maternally inherited, the MRCA of mtDNA represents the common ancestor of all women. The results show that Mitochondrial Eve lived about 200,000 years ago in Africa, which is one of the major evidences for the *Out of Africa* model.

The coalescent process also provides a flexible way to simulate genetic samples. The process can be split into two steps. In the first step, a coalescent tree is constructed according to a specified demographic model. During the coalesce process, random pairs (i, j) are chosen from all possible combinations and are assigned with a random waiting time that is exponentially distributed. The process continues backward in time until the MRCA of all genes is found. In the second step, the process goes forward in time. For each branch of the coalescent tree, the number of mutations is determined using a Poisson distribution with a parameter that is proportional to the length of the branch. Finally, starting from a random genotype of the MRCA, the mutation events are applied to each branch and the genotype of all genes on the coalescent tree is determined, which contains the sample we are interested.

The coalescent model can be extended to variable population size, population structure, and long sequences with recombination. However, due to the theoretical basis of the neutral theory, a coalescent process cannot model selection. It is then difficult to simulate the evolution of genetic diseases if a selection model is assumed for the disease allele. There have been a few approaches [15–18] that allow the simulation of selection using coalescent-like processes, but complex selection models such as multilocus selection model can still not be simulated.

A large number of programs are available to simulate genetic data using a coalescent approach. Although relatively limited in feature, Richard Hudson's ms program [19] remains one of the most popular programs. msHOT [20], SNPsim [21], COSI [22], CoaSim [23], SelSim [24], and others extend the algorithm to provide features such as varying recombination rates, gene conversion, complex demographic models, and a single-locus selection model. A survey of such software is provided in Liu et al [25].

1.2.10 Forward-Time Simulations

Forward-time population genetics simulation is simple in concept. Given a population with individuals of certain genotype, we evolve the population generation by generation, subject to various demographic and genetic forces such as population size change, mutation, selection, recombination, and migration. There are two key differences between forward-time and backward-time simulations:

1. Unlike coalescent-based approaches that simulate only individuals in a coalescent tree, forward-time simulations keep track of complete ancestral information. This gives forward simulations a wider application area if evolutionary processes themselves rather than their outcome are of interest [26] or if population-level properties are studied [27].
2. Because the coalescent theory is based on a neutral Wright–Fisher model, coalescent-based simulations can only simulate a random mating scheme with limited ability to simulate natural selection. In contrast, because there is no theoretical restriction, forward-time simulation approaches can simulate arbitrary nonrandom mating schemes and any genetic or environmental factors, so they can be used to simulate complex evolutionary processes that cannot be completely characterized by a Wright–Fisher model. This gives forward-time simulations a wider application area than coalescent-based simulations.

A number of forward-time simulation programs are available. If we exclude early applications developed primarily for teaching purposes, notable forward-time simulation programs include *easyPOP* [28], *Nemo* [29], *FreGene* [30], *GenomePop* [31], *ForwSim* [32], *ForSim* [33], and *SFS_CODE* [34], with an increasing list maintained at the "Other Simulation Tools" section of the simuPOP online cookbook. These programs are designed with specific applications and evolutionary scenarios in mind and excel for the purposes for which they were designed. A potentially large number of options are provided to allow users to choose from a number of stocked genetic models.

Although these programs can be used to simulate a large number of standard evolutionary processes, they are not flexible enough to be applied to problems outside their designed application areas. Researchers who work on complex evolutionary models or new application areas without existing simulation tools are usually forced to write their own software. In addition, because most of these programs are designed to simulate populations or samples resulting from certain evolutionary processes, they produce little output during the evolutionary processes. This makes it difficult to obtain, for example, the age of mutants in the simulated population when such information is needed.

In order to simulate and study a wide variety of evolutionary processes for different applications, we designed a general-purpose population genetics simulation program named simuPOP [35, 36]. In contrast to competing applications that use command-line options or configuration files to direct the execution of a limited number of predefined evolutionary scenarios, simuPOP provides users a large number of Python objects and functions, including individual, population, mating schemes, operators (objects that manipulate populations), and simulators to coordinate the evolutionary processes. This unique design makes simuPOP the best, and in many aspects the only software package for the simulation of complex evolutionary scenarios. This book uses simuPOP to simulate all examples. Readers who are interested in implementing their own simulations but are not familiar with simuPOP should refer to Appendix: "Forward-Time Simulations Using simuPOP" for an introduction to simuPOP.

1.3 STATISTICAL GENETICS AND GENETIC EPIDEMIOLOGY

1.3.1 Penetrance Models

A *genetic disease* is a disease caused by abnormal expression of one or more genes in a person, causing a clinically observed phenotype. One of the goals of genetic epidemiological studies is to locate these genes, usually called *disease predisposing loci* (DPL), so that we can detect, avoid, or develop treatments for these diseases. This process is called gene *mapping*.

Disease alleles are passed from parents to offspring, but do not always cause the phenotype (disease). The probability that certain genotype causes a phenotype is called the *penetrance* of the genotype. For example, if we assume that a disease is caused by a single locus, with one disease-causing allele a and one wild-type allele A, the penetrance for this disease is defined as the following:

$$\begin{aligned} f_0 &= \Pr(\text{affected} \mid AA) \\ f_1 &= \Pr(\text{affected} \mid Aa) \\ f_2 &= \Pr(\text{affected} \mid aa) \end{aligned}$$

The case f_0 is specifically termed the frequency of *phenocopies*, indicating individuals who are affected without carrying a disease-causing allele.

If f_i is either 0 or 1, the penetrance is called *full penetrance*. In the simplest case of one locus–two alleles, two disease alleles have to be present in a *recessive penetrance model* to cause the disease, whereas only one disease allele is needed in a *dominant penetrance model*. If the risk to be affected in a heterozygous person is half that of a person with two disease alleles, the penetrance is *additive*. *Multiplicative penetrance* can be similarly defined. These four basic models are listed in Table 1.3.

If the heterozygous phenotype is different from either of the homozygotes, these two alleles are called *codominant*. A simple example of codominant alleles is the human ABO gene that determines the blood types—A, B, AB, or O. In this case, the genotype at this locus fully determines the blood type of an individual. Among the three alleles, alleles A and B are dominant over allele O because allele A or B determines type A or B, irrespective of the presence of an O allele. Conversely, allele O is said

TABLE 1.3 Basic Penetrance Models.

	Genetic Models				
Genotype	General	Dominant	Recessive	Additive	Multiplicative
AA	f_0	0	0	0	0
Aa	f_1	1	0	s	s
aa	f_2	1	1	$2s$	s^2

TABLE 1.4 Penetrance at the ABO Locus.

	Phenotype			
Genotype	Type A	Type B	Type AB	Type O
A/A	1	0	0	0
A/B	0	0	1	0
A/O	1	0	0	0
B/B	0	1	0	0
B/O	0	1	0	0
O/O	0	0	0	1

to be recessive with respect to A or B. Alleles A and B are *codominant* because they are both expressed in the phenotype when present in the same genotype A/B (Table 1.4).

Multilocus penetrance models are much more complicated because the interaction between these disease susceptibility loci can be arbitrary. Only the simplest models have been studied, because of their mathematically tractability [37]. Assuming that the single-locus penetrance models are f_i, $i = 1, 2, \ldots$, a multiplicative multilocus penetrance model has overall penetrance $f = 1 - \prod_i (1 - f_i)$ and an additive model has overall penetrance $f = \max\left(1, \sum_i f_i\right)$. That is to say, assuming that all disease susceptibility loci follow an additive penetrance model $\left(0, s, 2s\right)$, an additive multilocus penetrance model has penetrance ns where n is the number of disease alleles at all loci.

Interaction between disease susceptibility loci can take different forms. Table 1.5 lists the penetrance of a two-locus epistatical penetrance model where the effect of dominant allele A at locus A is visible only when the individual also carries allele B at locus B. The model is symmetric in that the opposite is also true.

TABLE 1.5 An Example of Epistatical Penetrance Models.

	Locus B		
Locus A	BB	Bb	bb
AA	1	1	0
Aa	1	1	0
aa	0	0	0

1.3.2 Simple and Complex Genetic Diseases

If a genetic disease is caused by a single disease predisposing locus, it is called *monogenic*. Otherwise, it is called *polygenic*. Since monogenic diseases usually have a clear family inheritance pattern and high penetrance, they are easier to map. This may be why they are also called *simple diseases*. On the other hand, *polygenic* diseases usually do not obey the single-gene dominant or single-gene recessive Mendelian law. Their family inheritance traits are usually unclear because the disease can be caused by the joint effects of multiple genotypes and environmental factors. Because of the complexity in inheritance patterns, they are called *complex diseases*.

One hypothesis of the genetic basis of complex human disease is that much of the genetic variation underlying complex traits is derived from effects of common alleles [38–40]. This "Common Disease – Common Variant" hypothesis has important implications for disease mapping because common variants are relatively easy to detect by genetic association studies. More importantly, because common variants are more often in stronger linkage disequilibrium with their surrounding markers than rare variants, genetic associations are likely to exist for multiple markers if with the true disease locus. This makes it possible to use a fraction of *tagging markers* to perform genome-wide association studies.

However, despite the fact that hundreds of common variants have been identified for a number of complex diseases using genome-wide association studies [41, 42], they only explain a small proportion of genetic heritability of these diseases. It is therefore suspected that much, if not most, of the genetic risks of complex human diseases is due to a large number of rare variants with relatively high penetrance (the "Common Disease – Rare Variant" hypothesis) [43]. Because the statistical power of genome-wide association studies is substantially weaker in the detection of rare variants, new study designs and statistical methods are required to detect these variants.

1.3.3 Phenotypic, Allelic, and Locus Heterogeneity

All human genetic diseases could have been mapped and possibly treated if genetic diseases had homogeneous genotypes and phenotypes. Unfortunately, this is not the case. The same disease might develop at different ages and might show diverse characteristics (phenotype). The phenomenon that the same disease shows different features in different families or subgroups of patients is called *phenotypic heterogeneity*. It is often assumed that *phenotypic heterogeneity* is caused by different genetic factors, but there are cases in which the same known mutation causes variable symptoms or even multiple unrelated effects.

Genetic heterogeneity refers to the phenomenon of different mutations causing the same disease in different subgroups or families. More specifically, different alleles at the same locus (*allelic heterogeneity*) or at different loci (*locus heterogeneity*) can cause the same symptoms. A classical example of locus heterogeneity is given by mutations in the genes BRCA1 and BRCA2 that increase the risk of breast cancer. All these heterogeneity effects, plus incomplete penetrance that will be introduced below, pose prohibitive challenges in the mapping of complex human diseases. In the presence of allelic heterogeneity, when individuals are preferentially sampled due to having a specific disease, such as breast cancer, it is likely that specific mutant alleles will be in LD with different marker alleles. The result will be a loss of power to detect association with any of the marker alleles near the causal variant. For this reason, for example, association studies of breast cancer have not identified BRCA1 and BRCA2 despite the high penetrance of mutations in these genes.

1.3.4 Study Designs of Gene Mapping

The aim of gene mapping is to locate genetic variants on the genome that predispose to disease or some quantitative traits. There are two main approaches to gene mapping: linkage mapping in pedigrees and association mapping in the population. These two approaches differ in study design, statistical method, risk of false positive, and map resolution.

The basic principle of *linkage mapping* in pedigrees is the cosegregation of a phenotype and a genetic marker within families. If a marker is linked to a disease predisposing locus, it tends to be inherited together so that affected offspring tend to share the same chromosomal segments. Both nuclear families (two parents and their offspring) and large pedigrees (three or more generations) can be used for linkage studies, but one of the most common study designs uses affected sibpairs, namely, two affected siblings with their parents (if available).

A linkage test is a test of linkage between two markers, one of them can be disease status. This is done by testing if the recombination fraction between two markers is less than $\frac{1}{2}$. This hypothesis testing problem can be carried out using the likelihood ratio test. The LOD score is often used to assess the evidence for linkage, which is defined as

$$\mathrm{LOD}(\theta) = \log_{10} \frac{L(\mathrm{data} \mid \theta)}{L(\mathrm{data} \mid \theta = 1/2)},$$

where θ is a recombination frequency. Usually, we make a number of estimates of recombination frequency θ and estimate LOD score for each estimate. The estimate with the highest LOD score will be considered the best estimate. A computer program can optimize LOD from all θ between 0 and $1/2$. This test statistic differ from the standard likelihood rate test statistic by a constant factor (4.6). The range of θ is often truncated to be between 0 and 0.5, which leads to nonstandard testing conditions for this test (due to boundary constraints). A LOD score of 3 has been used as the threshold for linkage testing, which means the likelihood of observing the given pedigree if the two loci are unlinked is less than 0.001. If a frequentist approach to hypothesis testing is assumed, and the genome-wide level of significance is set to 0.05, allowing multiple testing from evaluating a genome-wide panel of markers, then a LOD score of 3.3 is sufficient for detecting genome-wide significant evidence for linkage.

Uncertainties in haplotypes (unknown phase), affection status (incomplete penetrance), and other factors can cloud the relationship between genotype and phenotype and reduce the power of this test. Even without these factors, the calculation of L (data $\mid \theta$) is a nontrivial task. The handling of large pedigrees, existence of loops in the pedigree (which occur when an allele can be transmitted through alternate paths, for instance, due to inbreeding), and large number of markers requires highly innovative computer algorithms. For large pedigrees with many markers, approximate methods such as MCMC (Markov Chain Monte Carlo) must be used. Reference [44] provides a thorough treatment of this topic.

Association studies look for a correlation between a specific variant and disease status or a quantitative trait in the population. For example, an association test for 2×2 contingency tables can be used to detect the association between a marker and disease status. Assume that we have a sample of M cases, and N controls. At each marker, the numbers of two alleles A and a in the case and control groups are counted as shown in Table 1.6. The χ^2 statistic for 2×2 contingency table is used to test the association between the marker and the disease status. This is the basic allele-based association test. More complex statistical tests include genotype-based association tests

TABLE 1.6 Illustration of the Case–Control Association Test

	Cases	Controls	Total
Allele A	a_{11}	a_{12}	a_1
Allele a	a_{21}	a_{22}	a_2
Total	$2M$	$2N$	$2M + 2N$

or trend tests. Reference [45] provides an excellent tutorial on statistical methods for population association studies.

REFERENCES

1. T. Strachan and A. P. Read, *Human Molecular Genetics*, Garland Science, 2003.
2. M. A. Jobling, M. Hurles, and C. Tyler-Smith, *Human Evolutionary Genetic*, Garland Science, 2004.
3. D. L. Hartl and A. G. Clark, *Principles of Population Genetics*, Sinauer Associates, Inc, 1997.
4. D. J. Balding, M. Bishop, and C. Cannings, *Handbook of Statistical Genetics*, John Wiley & Sons, Inc., 2003.
5. S. Myers, L. Bottolo, C. Freeman, G. McVean, and P. Donnelly, A fine-scale map of recombination rates and hotspots across the human genome. *Science*, 310(5746):321–324, 2005.
6. The neutral theory of molecular evolution, by Kimura M, Cambridge University Press http://www.amazon.com/Neutral-Theory-Molecular-Evolution/dp/0521317932/ref=sr_1_1?s=books&ie=UTF8&qid=1321048439&sr=1-1.
7. The causes of molecular evolution, by Gillespie, JH , Oxford University Press http://www.amazon.com/Causes-Molecular-Evolution-Oxford-Ecology/dp/0195092716/ref=sr_1_1?s=books&ie=UTF8&qid=1321048520&sr=1-1.
8. B. Charlesworth, Fundamental concepts in genetics: effective population size and patterns of molecular evolution and variation. *Nat Rev Genet*, 10(3): 195–205, 2009.
9. H. A. Orr, Fitness and its role in evolutionary genetics. *Nat Rev Genet*, 10(8):531–539, 2009.
10. J. K. Pritchard and M. Przeworski, Linkage disequilibrium in humans: models and data. *Am J Hum Genet*, 69(1):1–14, 2001.
11. M. Kimura and G. H. Weiss, The stepping stone model of population structure and the decrease of genetic correlation with distance. *Genetics*, 49(4):561–576, 1964.
12. P. Forster and S. Matsumura, Evolution: Did early humans go north or south? *Science*, 308(5724):965–966, 2005.
13. H. Liu, F. Prugnolle, A. Manica, and F. Balloux, A geographically explicit genetic model of worldwide human-settlement history. *Am J Hum Genet*, 79(2):230–237, 2006.
14. J. F. C. Kingman, The coalescent. *Stoch. Process. Appl*, 13:235–248, 1982.
15. S. M. Krone and C. Neuhauser, Ancestral processes with selection. *Theor Popul Biol*, 51(3):210–237, 1997.

16. P. Donnelly and T. G. Kurtz, Genealogical processes for Fleming-Viot models with selection and recombination. *Ann Appl Probab*, 9:1091–1148, 1999.
17. P. Fearnhead, Ancestral processes for non-neutral models of complex diseases. *Theor Popul Biol*, 63(2):115–130, 2003.
18. G. Coop and R. C. Griffiths, Ancestral inference on gene trees under selection. *Theor Popul Biol*, 66(3):219–232, 2004.
19. R. R. Hudson, Generating samples under a Wright-Fisher neutral model of genetic variation. *Bioinformatics*, 18(2):337–338, 2002.
20. G. Hellenthal and M. Stephens, mshot: modifying Hudson's ms simulator to incorporate crossover and gene conversion hotspots. *Bioinformatics*, 23(4):520–521, 2007.
21. D. Posada and C. Wiuf, Simulating haplotype blocks in the human genome. *Bioinformatics*, 19(2):289–290, 2003.
22. S. F. Schaffner, C. Foo, S. Gabriel, D. Reich, M. J. Daly, and D. Altshuler, Calibrating a coalescent simulation of human genome sequence variation. *Genome Res*, 15(11):1576–1583, 2005.
23. T. Mailund, M. H. Schierup, C. N. S. Pedersen, P. J. M. Mechlenborg, J. N. Madsen, and L. Schauser, CoaSim: a flexible environment for simulating genetic data under coalescent models. *BMC Bioinformatics*, 6:252, 2005.
24. C. C. A. Spencer and G. Coop, SelSim: a program to simulate population genetic data with natural selection and recombination. *Bioinformatics*, 20(18):3673–3675, 2004.
25. Y. Liu, G. Athanasiadis, and M. E. Weale, A survey of genetic simulation software for population and epidemiological studies. *Hum Genom*, 3(1): 79–86, 2008.
26. F. Calafell, E. L. Grigorenko, A. A. Chikanian, and K. K. Kidd, Haplotype evolution and linkage disequilibrium: a simulation study. *Hum Hered*, 51(1–2):85–96, 2001.
27. F. Balloux and J. Goudet, Statistical properties of population differentiation estimators under stepwise mutation in a finite island model. *Mol Ecol*, 11(4): 771–783, 2002.
28. F. Balloux, Easypop (version 1.7): a computer program for population genetics simulations. *J Hered*, 92(3):301–302, 2001.
29. F. Guillaume and J. Rougemont, Nemo: an evolutionary and population genetics programming framework. *Bioinformatics*, 22(20):2556–2557, 2006.
30. M. Chadeau-Hyam, C. J. Hoggart, P. F. O'Reilly, J. C. Whittaker, M. De Iorio, and D. J Balding, Fregene: simulation of realistic sequence-level data in populations and ascertained samples. *BMC Bioinformatics*, 9:364, 2008.
31. A. Carvajal-RodrÃguez, Genomepop: a program to simulate genomes in populations. *BMC Bioinformatics*, 9:223, 2008.

32. B. Padhukasahasram, P. Marjoram, J. D. Wall, C. D. Bustamante, and M. Nordborg, Exploring population genetic models with recombination using efficient forward-time simulations. *Genetics*, 178(4):2417–2427, 2008.
33. B. W. Lambert, J. D. Terwilliger, and K. M. Weiss, Forsim: a tool for exploring the genetic architecture of complex traits with controlled truth. *Bioinformatics*, 24(16):1821–1822, 2008.
34. R. D Hernandez, A flexible forward simulator for populations subject to selection and demography. *Bioinformatics*, 24(23):2786–2787, 2008.
35. B. Peng and M. Kimmel, simuPOP: a forward-time population genetics simulation environment. *Bioinformatics*, 21(18):3686–3687, 2005.
36. B. Peng and C. I. Amos, Forward-time simulations of non-random mating populations using simuPOP. *Bioinformatics*, 24(11):1408–1409, 2008.
37. N. Risch, Linkage strategies for genetically complex traits: II. The power of affected relative pairs. *Am J Hum Genet*, 46(2):229–241, 1990.
38. E. S. Lander and N. J. Schork, Genetic dissection of complex traits. *Science*, 265(5181):2037–2048, 1994.
39. N. Risch and K. Merikangas, The future of genetic studies of complex human diseases. *Science*, 273(5281):1516–1517, 1996.
40. E. S. Lander, The new genomics: global views of biology. *Science*, 274(5287):536–539, 1996.
41. R. McPherson, A. Pertsemlidis, N. Kavaslar, A. Stewart, R. Roberts, D. R. Cox, D. A. Hinds, L. A. Pennacchio, A. Tybjaerg-Hansen, A. R. Folsom, E. Boerwinkle, H. H. Hobbs, and J. C. Cohen, A common allele on chromosome 9 associated with coronary heart disease. *Science*, 316(5830):1488–1491, 2007.
42. C. I. Amos, Successful design and conduct of genome-wide association studies. *Hum Mol Genet*, 16(2):R220–R225, 2007.
43. N. S. Fearnhead, B. Winney, and W. F. Bodmer, Rare variant hypothesis for multifactorial inheritance: susceptibility to colorectal adenomas as a model. *Cell Cycle*, 4(4):521–525, 2005.
44. Analysis of Human Genetic Linkage, by Jurg Ott, The Jons Hopkins University Press (http://www.amazon.com/Analysis-Human-Genetic-Linkage-Ott/dp/0801861403).
45. D. J Balding, A tutorial on statistical methods for population association studies. *Nat Rev Genet*, 7(10):781–791, Oct 2006.

CHAPTER 2

SIMULATION OF POPULATION GENETICS MODELS

This chapter uses simuPOP to simulate a number of standard population genetics models. The goal of these simulations is to demonstrate how to use various simuPOP features to simulate genetic factors such as mutation, selection, and recombination. Because detailed descriptions of these models are widely available in textbooks such as *Principles of Population Genetics* [1], we describe these models and their theoretical properties briefly, only as a way to motivate our simulations. Although simulations in this chapter are confirmatory in nature, they could be used to form the basis of more complex evolutionary scenarios in which more than one genetic factor would be applied.

2.1 RANDOM GENETIC DRIFT

The Wright–Fisher model is a model of random genetic drift with binomial sampling, which is characterized by the independent binomial sampling of parents with replacement. A standard diploid Wright–Fisher model assumes a constant population of N individuals ($2N$ chromosomes). Because offspring chromosomes are chosen independently from an infinite pool of

Forward-time Population Genetics Simulations: Methods, Implementation, and Applications,
Bo Peng, Marek Kimmel, and Christopher I. Amos.

gametes, this process is equivalent to a haploid Wright–Fisher model with $2N$ individuals.

2.1.1 Dynamics of Allele Frequency and Heterozygosity

The basic property of random genetic drift is the random drift of allele frequency due to binomial sampling. Assuming that there are no additional genetic forces such as mutation and natural selection, let Y_n be the number and $X_n = \frac{Y_n}{2N}$ be the frequency of allele 1 of a diallelic marker at generation n, the distribution of the number of alleles at generation n follows a binomial distribution with parameters $2N$ and $x_{n-1} = \frac{y_{n-1}}{2N}$,

$$\Pr(Y_n \mid Y_{n-1} = y_{n-1}) = \text{Binomial}\left(2N, \frac{y_{n-1}}{2N}\right). \tag{2.1}$$

Assuming $x_1 = p$ is the starting allele frequency, the expected allele frequency $E(X_n)$ keeps constant because $E(X_n \mid X_{n-1}) = X_{n-1}$. In addition, because the probability that an individual being homozygous in the next generation is

$$F_{t+1} = \frac{1}{2N} + \left(1 - \frac{1}{2N}\right) F_t,$$

the probability of an individual being heterozygous is

$$H_t = 1 - F_t = \left(1 - \frac{1}{2N}\right) H_{t-1} = \left(1 - \frac{1}{2N}\right)^t H_0. \tag{2.2}$$

That is to say, the proportion of heterozygotes in this population is expected to decay exponentially.

■ **EXAMPLE 2.1**

Let us simulate a basic Wright–Fisher process of random genetic drift with $N = 100$ and $p = 0.5$. Although we can run a large number of simulations sequentially or simultaneously by running replicates of the same population (as shown in Source Code A.25), this example evolves a large population with 100 subpopulations, each with 100 individuals. Because a random mating scheme evolves each subpopulation separately in the absence of migration, such a simulation effectively evolves `100` populations independently.

This example calculates the frequency of allele 1 and the proportion of heterozygotes of each subpopulation at generations 0, 20, ..., and 80, and stores the results in variables subPop[sp]['alleleFreq'] and subPop[sp]['heteroFreq'] where sp is a subpopulation index. It then uses a PyEval operator to calculate and output the average allele and heterozygote frequencies across all subpopulations using a relatively complex Python expression. In addition to generation number and mean observed allele frequency and heterozygote frequency, this expression also calculates the expected heterozygosity using formula 2.2 with $H_0 = 2pq = 0.5$.

As we can see from the output of this example, although the mean frequency in 100 subpopulations remains close to 0.5, heterozygosity in each subpopulation decreases (although allele frequencies deviate from the initial frequency in different directions), resulting in a decreased average proportion of heterozygotes in these subpopulations.

SOURCE CODE 2.1 Decay of Homozygosity Due to Random Genetic Drift

```
>>> import simuPOP as sim
>>> # [100]*100 means a population with 100 subpopulations, each of size 100.
>>> pop = sim.Population([100]*100, loci=1)
>>> pop.evolve(
...     initOps=[
...         sim.InitSex(),
...         sim.InitGenotype(freq=[0.5, 0.5]),
...         sim.PyOutput('gen:   mean freq   mean Ht (expected Ht)\gn')
...         ],
...     preOps=[
...         # Statistics in subpopulations are by default not calculated.
...         # This is changed by specifying variables with a '_sp' suffix.
...         sim.Stat(alleleFreq=0, heteroFreq=0,
...             vars=['alleleFreq_sp', 'heteroFreq_sp'], step=20),
...         # Variables in subpopulations are stored in dictionaries
...         # subPop[subPopID] where subPopID can be virtual.
...         sim.PyEval(r'"%2d:    %.4f      %.4f (%.4f)\n" % (gen,'
...             'sum([subPop[x]["alleleFreq"][0][1] for x in range(100)])/100.,'
...             'sum([subPop[x]["heteroFreq"][0] for x in range(100)])/100.,'
...             '0.5*(1-1/200.)**gen)', step=20)
...     ],
...     matingScheme=sim.RandomMating(),
...     gen=100
... )
gen:   mean freq   mean Ht (expected Ht)
 0:    0.4982      0.4951 (0.5000)
20:    0.5017      0.4692 (0.4523)
40:    0.5260      0.4182 (0.4092)
60:    0.5343      0.3744 (0.3701)
80:    0.5272      0.3303 (0.3348)
100
```

2.1.2 Persistence Time

Because X_n is bounded by frequencies 0 and 1, alleles in a standard Wright–Fisher process will eventually become fixed or lost. Assuming an initial allele frequency p, the mean time (in generation) until an allele is fixed conditioning on it is eventually fixed at

$$t_1(p) = -4N\left(\frac{1-p}{p}\right)\ln(1-p). \tag{2.3}$$

Similarly, the mean time to loss until an allele is lost is

$$t_0(p) = -4N\left(\frac{p}{1-p}\right)\ln p. \tag{2.4}$$

Because the probability that an allele gets lost (instead of fixed) equals its initial allele frequency p, the mean persistence time of an allele is

$$t(p) = pt_1(p) + (1-p)t_0(p) = -4N\left[(1-p)\ln(1-p) + p\ln(p)\right]. \tag{2.5}$$

■ EXAMPLE 2.2

Using a terminator that terminates the evolution of a population after an allele gets lost or fixed, this example simulates the persistence time of `500` populations with $N = 100$ and $p = 0.4$. It initializes allele 0 with a frequency of 0.4 and uses a `RandomSelection` mating scheme to simulate the random binomial selection of gametes in a population with $2N$ chromosomes. This mating scheme is almost identical to a `RandomMating` mating scheme in a population with N individuals except that it ignores individual sex and chooses gametes from a pool of $2N$ chromosomes randomly.

At the end of each generation, a `Stat` operator is used to calculate the frequency of alleles at locus 0. A `Terminator` is then used to terminate the evolution of a population if variable `alleleFreq[0][0]` in its local namespace is either `0` or `1`. Function `Population.evolve` returns after all populations have been terminated and returns the number of generations each population has evolved.

The rest of the example goes through each population and determines whether or not the allele is lost or fixed by checking its frequency at the end of evolution. The last `print` statement outputs the observed mean persistence times for all populations and populations with fixed or lost alleles and compares them with their theoretical estimates calculated from Equations 2.3–2.5.

SOURCE CODE 2.2 Absorption Time and Time to Fixation

```
>>> import simuPOP as sim
>>> from math import log
>>> N = 100
>>> p = 0.4
>>> pop = sim.Population(2*N, ploidy=1, loci=1)
>>> # Use a simulator to simulate 500 populations simultaneously.
>>> simu = sim.Simulator(pop, rep=500)
>>> gens = simu.evolve(
...     initOps=sim.InitGenotype(prop=[p, 1-p]),
...     # A RandomSelection mating scheme choose parents randomly regardless
...     # of sex and copy parental genotype to offspring directly.
...     matingScheme=sim.RandomSelection(),
...     postOps=[
...         # calculate allele frequency at locus 0
...         sim.Stat(alleleFreq=0),
...         # and terminate the evolution of a population if it has no
...         # allele 0 or 1 at locus 0.
...         sim.TerminateIf('alleleFreq[0][0] in (0, 1)'),
...     ],
... )
>>> # find out populations with or without allele 1
>>> gen_lost = []
>>> gen_fixed = []
>>> for gen,pop in zip(gens, simu.populations()):
...     if pop.dvars().alleleFreq[0][0] == 0:
...         gen_lost.append(gen)
...     else:
...         gen_fixed.append(gen)
...
>>> print('''\nMean persistence time: %.2f (expected: %.2f)
... Lost pops: %d (expected: %.1f), Mean persistence time: %.2f (expected: %.2f)
... Fixed pops: %d (expected: %.1f), Mean persistence time: %.2f (expected: %.2f)'''\
... % (float(sum(gens)) / len(gens), -4*N*((1-p)*log(1-p) + p*log(p)),
... len(gen_fixed), 500*p, float(sum(gen_fixed)) / len(gen_fixed),
... -4*N*(1-p)/p*log(1-p), len(gen_lost), 500*(1-p),
... float(sum(gen_lost)) / len(gen_lost), -4*N*p/(1-p)*log(p)))

Mean persistence time: 268.42 (expected: 269.20)
Lost pops: 195 (expected: 200.0), Mean persistence time: 308.99 (expected: 306.50)
Fixed pops: 305 (expected: 300.0), Mean persistence time: 242.48 (expected: 244.34)
>>>
```

2.2 DEMOGRAPHIC MODELS

Human and other populations have experienced various complex demographic events such as population splitting and expansion during evolution. Because the demographic history of a population has profound impact on the genetic composition of the population, it is important to incorporate realistic demographic histories in the simulation of human populations. Being a forward-time simulator, simuPOP requires the specification of the exact number and sizes of subpopulations at every generation. Once one has a clear idea which demographic model to use, it is generally easy to implement it using the mechanism described in this chapter.

2.2.1 The Bottleneck Effect

Because the variance of the binomial distribution of Equation 2.1 increases with smaller population size N, the impact of random genetic drift is more rapid in small populations than in large populations. If the population size grows rapidly after a period of small size, the increased population size tends to decrease the force of subsequent genetic drift and therefore freeze the impact of dramatic changes that occurred before population expansion. This phenomenon, called a *bottleneck effect,* plays an important role in the evolution of many human populations.

■ EXAMPLE 2.3

This example simulates a demographic model where a population of size 1000 is evolved for 40 generations before experiencing a bottleneck comprising 20 individuals for 10 generations. The population size then rebounds to 1000 individuals and maintains a constant population size for another 50 generations. This demographic model is implemented using a function `demo` that returns a population size of 1000 or 20 according to the passed generation number.

In order to study the impact of this bottleneck on the evolution of alleles, we trace the frequency of an allele in five replicate simulations. Instead of reporting allele frequencies for each replicate, this example uses a `VarPlotter` operator to plot the trajectories of the frequency of allele `0` during evolution. This operator is defined in a utility module `simuPOP.plotter`. In addition to a working R environment, a Python module `rpy` (`http://rpy.sourceforge.net`) that provides a link between Python and R is needed to make use of this module.

Operator `VarPlotter` evaluates expression `alleleFreq[0][0]` for each replicate and plots the current and historical values of this expression at every 10 generations (parameter `update=10`). Because the figures will be updated at every 10 generations, an animation will be displayed during the evolution of this population. The figures are displayed on the screen and are saved to files `ch3_bottleneck_x.pdf` ($x = 10, \ldots, 100$) because of the use of parameter `saveAs`.

A visualization operator usually calls more than one R functions such as `par`, `plot`, and `lines` to plot a figure. When an additional keyword parameter is provided to this operator, it will be passed to R functions if the parameter name is prefixed with a function name. For example, parameters `plot_ylim=[0,1]` and `plot_ylab='Allele Frequency'` pass parameters `ylim=[0,1]` and `ylab='Allele Frequency'` to function `plot` to customize the plots.

As we can see from the trajectory of allele frequencies at generation 100 (Figure 2.1), because of a relatively large population size (`1000`), these alleles have similar allele frequencies in the first 40 generations. Frequencies of these alleles change rapidly due to strong genetic drift during the bottleneck period, but stabilize again after the population size rebounds to 1000 individuals.

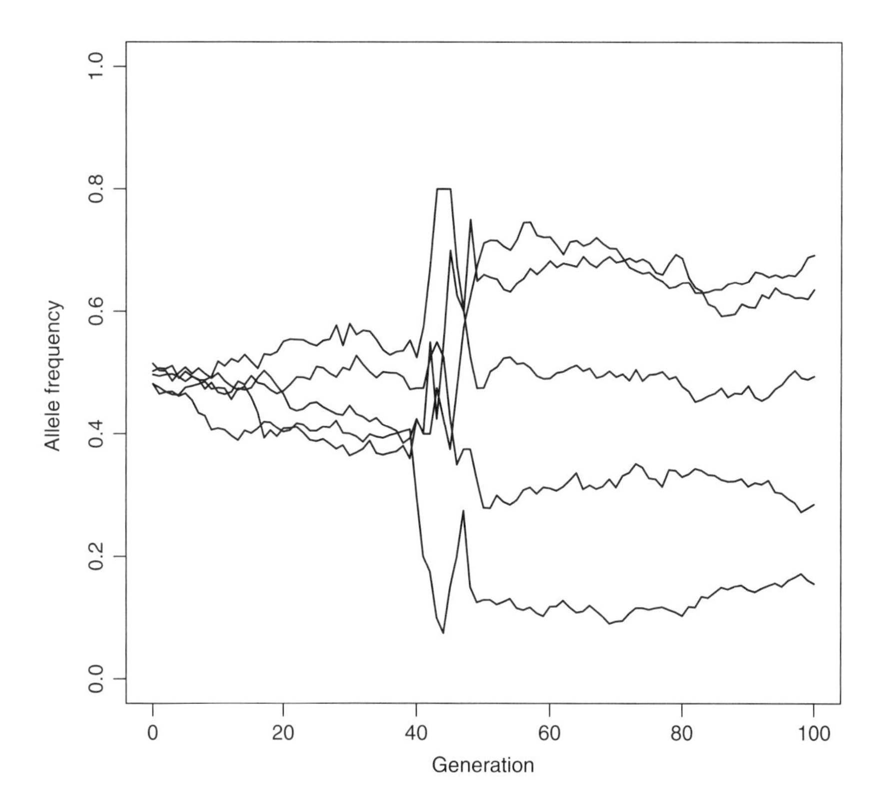

FIGURE 2.1 Impact of population bottleneck. Simulation of genetic drift in five replicate populations starting with initial allele frequency of 0.5 over a period of 100 generations. The population size is kept constant at 1000 individuals until generation 40, at which point the population size is reduced to 20 individuals. The population size then rebounds to 1000 individuals at generation 49 and remains at this population size for the remainder of the simulation.

SOURCE CODE 2.3 Demonstration of a Bottleneck Effect

```
import simuPOP as sim
from simuPOP.plotter import VarPlotter
def demo(gen):
    # split subpopulation 0 at generation 10 and 20
    if gen >=40 and gen < 50:
        return 20
    else:
        return 1000

pop = sim.Population(size=1000, loci=1)
simu = sim.Simulator(pop, rep=5)
simu.evolve(
    initOps=[
        sim.InitSex(),
        sim.InitGenotype(freq=[0.5, 0.5])
    ],
    matingScheme=sim.RandomMating(subPopSize=demo),
    postOps=[
        sim.Stat(alleleFreq=0),
        VarPlotter('alleleFreq[0][0]', update=10,
            saveAs='Figures/bottleneck.pdf',
            plot_ylim=[0,1], plot_ylab='Allele Frequency',
            lines_col='black', lines_lwd=1.5)
    ],
    gen=101
)
```

2.3 MUTATION

There are many mutation models for different types of genetic markers and models, and the definition of a *mutation rate* can differ from model to model. This section demonstrates how to use simuPOP mutators (mutation operators) to simulate a number of models for single-nucleotide polymorphism (SNP) and microsatellite markers.

2.3.1 A Diallelic Mutation Model

A mutation model with n allelic states can be defined with a *mutation rate matrix* $\left(p_{ij}\right)_{n\times n}$ where p_{ij} is the probability that an allele i mutates to allele j per locus per generation. If we start with a specific allele at a locus, this matrix determines the distribution of allelic states at the next generation. Running this process for many steps creates a Markov chain because the next state depends on only the current state and on the mutation matrix. Therefore, a mutation rate matrix is a *transition matrix* of a Markov process. Another concept, called a *substitution matrix,* is often used in the theoretical analysis of these models. A substitution matrix is defined to be

$Q = R - I$ where I is the identity matrix, so the rows of such a matrix sum to zero.

A mutation rate matrix for a diallelic model with a wild-type allele A and a disease allele a has two free variables, namely, a forward mutation rate u (the probability that an allele A mutates to allele a at each generation) and a backward mutation rate v (the probability that an allele a mutates to allele A). The mutation rate matrix for this model is

$$\begin{pmatrix} u & 1-u \\ 1-v & v \end{pmatrix}.$$

Assuming that the frequency of allele A is p_{t-1} at generation $t-1$, its frequency at the next generation, if we do not consider the impact of genetic drift, would be

$$p_t = p_{t-1}(1-u) + v(1-p_{t-1}) \tag{2.6}$$

because $1-u$ of allele A will remain as allele A and v of allele a will be mutated to allele A. Mathematical manipulation of Equation 2.6 leads to

$$p_t - \frac{v}{u+v} = \left(p_0 - \frac{v}{u+v}\right)(1-u-v)^t, \tag{2.7}$$

which implies an equilibrium allele frequency $p = \frac{v}{u+v}$ if $t \to \infty$. Of course, if $u = 0$ or $v = 0$, the equilibrium allele frequency will be 0 or 1 because one of the alleles will mutate to the other.

■ EXAMPLE 2.4

simuPOP provides a mutation operator `SNPMutator(u,v)` to simulate a mutation model with mutation rate matrix

$$\begin{pmatrix} u & 1-u \\ 1-v & v \end{pmatrix}.$$

This mutator is named `SNPMutator` because it is usually applied to single-nucleotide polymorphism markers that has two allelic states.

This example evolves three populations of size 10,000 for 500 generations. Instead of applying the same operator to all populations, this example uses three `InitGenotype` operators to initialize them with different starting allele frequencies 0.8, 0.5, and 0.2, respectively. The trick here is the use of parameter `reps`, which accepts one or more indices of populations on which an operator is applicable. Negative indices are also acceptable.

For example, operator `PyOutput` in this example outputs a newline string after the last population has been evolved (`reps=-1`).

A `SNPMutator` is used to mutate between alleles 0 and 1 with mutation rates $u = 10^{-2}$ and $v = 10^{-3}$. As we can see from the output, allele frequencies at these populations gradually approach and then oscillate around (due to genetic drift) an equilibrium allele frequency of $\frac{v}{u+v} = 9.1\%$.

SOURCE CODE 2.4 Diallelic Mutation Model

```
>>> import simuPOP as sim
>>> from math import log
>>> pop = sim.Population(size=10000, loci=1)
>>> simu = sim.Simulator(pop, rep=3)
>>> simu.evolve(
...     initOps=[
...         sim.InitSex(),
...         # Because of the use of parameter reps, these operators are
...         # applied to different populations.
...         sim.InitGenotype(freq=(0.2, 0.8), reps=0),
...         sim.InitGenotype(freq=(0.5, 0.5), reps=1),
...         sim.InitGenotype(freq=(0.8, 0.2), reps=2)],
...     preOps=sim.SNPMutator(u=0.01, v=0.001),
...     matingScheme=sim.RandomMating(),
...     postOps=[
...         sim.Stat(alleleFreq=0, step=100),
...         sim.PyEval('gen', reps=0, step=100),
...         sim.PyEval(r"'\t%.3f' % alleleFreq[0][0]", step=100),
...         sim.PyOutput('\n', reps=-1, step=100)
...     ],
...     gen=500
... )
0 0.200 0.499 0.793
100 0.133 0.196 0.340
200 0.115 0.130 0.189
300 0.099 0.109 0.111
400 0.099 0.088 0.114
(500, 500, 500)
```

2.3.2 Multiallelic Mutation Models

A multiallelic mutation model mutates between multiple alleles at a locus. A *stepwise mutation model* is usually used to model the mutation of microsatellite markers. The basic form of this model, namely, a symmetrical stepwise mutation model, assumes that the entire sequence of allelic states can be described by integers and that, if an allele changes state by mutation, it moves either one step in the positive direction or one step in the negative direction in the allele space. In comparison, a k-allele mutation model assumes that there are k alleles at a locus and a mutation event mutates an allele to any other allele. A k-allele model with a sufficiently large k

can be used to approximate an infinite allele model because a mutant in such a model has a high probability to be a new mutant. Stepwise models have been used to describe mutations among repetitive DNA sequences such as microsatellites, while k-allele models would be appropriate more generically.

■ EXAMPLE 2.5

This example evolves two populations each with 10 subpopulations. Individuals in both populations are initialized with random sex and alleles 100 at the beginning of the evolutionary process. We choose an initial allele of 20 because we assume that there are already 20 tandem repeats at the microsatellite locus.

During evolution, a stepwise mutation model with mutation rate $\mu = 10^{-4}$ is applied to the first population, and a k-allele mutation model with $k = 10{,}000$ and with the same mutation rate is applied to the second population. An operator `Stat` is used to calculate allele frequency at all subpopulations, which are used to calculate the mean number of different alleles contained in the populations. Because alleles are more likely to mutate to an existing allelic state in a stepwise mutation model than in a k-allele model, the average number of distinct alleles in the first population is smaller than that in the second population, as predicted in Ref. [2].

SOURCE CODE 2.5 *k*-Allele and Stepwise Mutation Models

```
>>> import simuOpt
>>> simuOpt.setOptions(quiet=True, alleleType='long')
>>> import simuPOP as sim
>>> pop = sim.Population(size=[2500]*10, loci=1)
>>> simu = sim.Simulator(pop, rep=2)
>>> simu.evolve(
...     initOps=[
...         sim.InitSex(),
...         sim.InitGenotype(genotype=20),
...     ],
...     preOps=[
...         sim.StepwiseMutator(rates=0.0001, reps=0),
...         sim.KAlleleMutator(k=10000, rates=0.0001, reps=1),
...     ],
...     matingScheme=sim.RandomMating(),
...     postOps=[
...         # Use vars=['alleleFreq_sp'] to calculate allele frequency for
...         # each subpopulation
...         sim.Stat(alleleFreq=0, vars=['alleleFreq_sp'], step=200),
...         sim.PyEval('gen', step=200, reps=0),
...         sim.PyEval(r"'\t%.2f' % (sum([len(subPop[x]['alleleFreq'][0]) "
...             "for x in range(10)])/10.)", step=200),
...         sim.PyOutput('\n', reps=-1, step=200)
...     ],
```

```
...      gen=601
... )
0 1.30 1.40
200 2.60 5.30
400 2.70 6.10
600 2.80 6.60
(601, 601)
```

2.4 MIGRATION

Subpopulations evolve separately and tend to become more and more genetically distinct due to the impact of random genetic drift. Migration allows the exchange of individuals, and thus genotypes, between subpopulations. It glues subpopulations together and limits the genetic divergence between subpopulations.

2.4.1 An Island Model of Migration

In an island model of migration, a large population is split into many subpopulations like islands in an archipelago. Migration in a basic island model is assumed to occur evenly between all islands so that the probability that a randomly chosen allele in any subpopulation comes from a migrant is the same across islands. Under some simplified conditions (e.g., no genetic drift), the frequency of an allele in one island at generation t equals to

$$p_t - \bar{p} = (p_0 - \bar{p})(1 - m)^t ,$$

where m is the migration rate, p_0 is the initial frequency, and $\bar{p}$ is the average frequency across all subpopulations. Obviously, all islands will have the same allele frequency $\bar{p}$ when t approaches infinity.

■ EXAMPLE 2.6

simuPOP provides an operator `Migrator` to migrate individuals between subpopulations. It uses a migration matrix (m_{ij}) where m_{ij} is the probability that an individual moves from subpopulation i to subpopulation j at each generation. Diagonal items of this matrix, namely, the probability that an individual stays in his or her own subpopulation, can be left unspecified because they are determined automatically from $m_{ii} = 1 - \sum_{j \neq i} m_{ij}$. Because operator `Migrator` needs to record the destination subpopulation of each individual before it moves individuals, it requires an information field `migrator_to` from the populations to which. Note that this operator can also migrate individuals by proportion and by individual count, and can migrate to new subpopulations, which effectively creates subpopulations with migrants.

This example creates two populations, each with three subpopulations of 10,000 individuals. The frequencies of allele 0 for these three subpopulations are initialized to 0.2, 0.3, and 0.5, respectively. At the beginning of each generation, individuals in the first population migrate between three subpopulations following an island model with migration rate 0.01. For comparison purposes, no migration is allowed for the second population.

This example calculates allele frequencies in subpopulations and uses a F_{ST} statistic, a measure of population differentiation based on genetic differences between populations, to quantify the genetic diversity among three subpopulations. F_{ST} is calculated as the correlation of randomly chosen alleles within the same subpopulation relative to that found in the entire population. It should be close to 0 if populations are genetically close due to free interbreeding.

We evolve the population for 201 generations and output F_{ST} and allele frequencies in the three subpopulations at every 50 generations. As we can see from the output, allele frequencies of the first population approach an equilibrium frequency of 0.33, whereas those in the second populations remain distinct, which are also reflected in the dynamics of the F_{ST} statistics of the two populations.

SOURCE CODE 2.6 An Island Model of Migration

```
>>> import simuPOP as sim
>>> from simuPOP.utils import migrIslandRates
>>> p = [0.2, 0.3, 0.5]
>>> pop = sim.Population(size=[10000]*3, loci=1, infoFields='migrate_to')
>>> simu = sim.Simulator(pop, rep=2)
>>> simu.evolve(
...     initOps=[sim.InitSex()] +
...         [sim.InitGenotype(prop=[p[i], 1-p[i]], subPops=i) for i in range(3)],
...     preOps=sim.Migrator(rate=migrIslandRates(0.01, 3), reps=0),
...     matingScheme=sim.RandomMating(),
...     postOps=[
...         sim.Stat(alleleFreq=0, structure=0, vars='alleleFreq_sp', step=50),
...         sim.PyEval("'Fst=%.3f (%s)' % (F_st, ', '.join(['%.2f' % "
...             "subPop[x]['alleleFreq'][0][0] for x in range(3)]))",
...             step=50),
...         sim.PyOutput('\n', reps=-1, step=50),
...     ],
...     gen=201
... )
Fst=0.101 (0.20, 0.30, 0.50) Fst=0.101 (0.20, 0.30, 0.50)
Fst=0.024 (0.28, 0.31, 0.42) Fst=0.113 (0.20, 0.27, 0.51)
Fst=0.011 (0.30, 0.34, 0.40) Fst=0.085 (0.24, 0.24, 0.48)
Fst=0.004 (0.38, 0.33, 0.38) Fst=0.141 (0.22, 0.23, 0.54)
Fst=0.001 (0.39, 0.36, 0.37) Fst=0.166 (0.19, 0.22, 0.55)
(201, 201)
```

2.5 RECOMBINATION AND LINKAGE DISEQUILIBRIUM

Genetic recombination breaks existing parental chromosomes and introduces new recombinant chromosomes to the population. If two loci are linked (located on the same chromosome), recombination tends to reduce the linkage disequilibrium between these two loci. The rate at which linkage disequilibrium decays follows Equation 1.11 (see Example 2.7). Although it is relatively easy to model recombination between two adjacent loci, the impact of recombination on multiple linked loci, especially when natural selection is involved, is a complicated issue.

■ EXAMPLE 2.7

This example simulates a population of 10,000 individuals. Each individual has a chromosome with 10 loci. These loci are arranged in two groups, at locations 0, 1, 2, 3, 4, 10, 11, 12, 13, 14. The population is initialized with random sex and two haplotypes, one with all 0 alleles and another with all 1 alleles. The linkage disequilibrium D are therefore 0.25 between all loci at the beginning of this evolutionary process.

We evolve the population for 100 generations and use a `Recombinator` operator to recombine parental chromosomes before they are transmitted to offspring. Instead of specifying recombination rates between adjacent loci explicitly, this example specifies recombination rates as the product of loci distance and a recombination intensity 5×10^{-4}. For example, the recombination rate between loci 1 and 2 is 5×10^{-4} because they are located 1 unit apart, and the recombination rate between loci 4 and 5 is 2.5×10^{-3} because they are 5 units away from each other.

We calculate and display linkage disequilibrium values between loci $(1, 2)$, $(4, 5)$, and $(0, 9)$, along with expected linkage disequilibrium between loci (1, 2) and (4, 5). As we can see from the output of the example, the linkage disequilibrium values between loci 1 and 2 decay more slowly than values between loci 4 and 5 because of stronger recombination between these loci 4 and 5.

SOURCE CODE 2.7 Recombination Between Three Loci

```
>>> import simuPOP as sim
>>> pop = sim.Population(size=10000, loci=10, lociPos=range(5) + range(10, 15))
>>> pop.evolve(
...     initOps=[
...         sim.InitSex(),
...         sim.InitGenotype(haplotypes=[[0]*10,[1]*10]),
...     ],
...     matingScheme=sim.RandomMating(ops=sim.Recombinator(intensity=0.0005)),
```

```
...     postOps=[
...         sim.Stat(LD=[[1,2],[4,5],[8,9],[0,9]], step=10),
...         sim.PyEval(r"'gen=%d\tLD12=%.3f (%.3f)\tLD45=%.3f (%.3f)\tLD09=%.3f\n'%"
...             "(gen, LD[1][2], 0.25*0.9995**(gen+1), LD[4][5],"
...             "0.25*0.9975**(gen+1),LD[0][9])", step=10)
...     ],
...     gen=100
... )
gen=0 LD12=0.250 (0.250) LD45=0.249 (0.249) LD09=0.248
gen=10 LD12=0.249 (0.249) LD45=0.242 (0.243) LD09=0.232
gen=20 LD12=0.248 (0.247) LD45=0.234 (0.237) LD09=0.213
gen=30 LD12=0.247 (0.246) LD45=0.225 (0.231) LD09=0.201
gen=40 LD12=0.246 (0.245) LD45=0.216 (0.226) LD09=0.189
gen=50 LD12=0.246 (0.244) LD45=0.214 (0.220) LD09=0.178
gen=60 LD12=0.245 (0.242) LD45=0.208 (0.215) LD09=0.166
gen=70 LD12=0.246 (0.241) LD45=0.202 (0.209) LD09=0.156
gen=80 LD12=0.245 (0.240) LD45=0.193 (0.204) LD09=0.147
gen=90 LD12=0.245 (0.239) LD45=0.186 (0.199) LD09=0.136
100
```

2.6 NATURAL SELECTION

Simulation of natural selection can be achieved in two ways: selection of parents using *relative fitness* values and selection of offspring using *absolute fitness* values. The first method is easier to use and is more efficient, and the second method can be useful in modeling certain evolutionary processes when multiple offspring are produced and selected.

Selection of parents is performed during mating. In this case, all parents are assigned a fitness value $f_i \geq 0, i = 1, ..., N$. During each mating event, parents are chosen at a probability that is proportional to their fitness values. More specifically, in the standard Wright–Fisher model with natural selection, the probability that individual k is selected for mating is $\frac{f_k}{\sum_{i=1}^{N} f_i}$ at each mating event.

Another mechanism for natural selection is performed during the production of offspring. Using this method, a fitness value is assigned for each offspring, which is interpreted directly as the survival probability of this individual. For example, if an individual has fitness 0.9, it will have a probability of 0.1 to be discarded, regardless of fitness values of other offspring. Because most theoretical models use relative fitness and a large number of offspring will be discarded if small fitness values are used, this method is used less frequently than the selection of parents method.

2.6.1 Single-Locus Diallelic Selection Models

Assuming that the relative fitness of three genotypes AA, Aa, and aa at a locus are 1, $1 - hs$, and $1 - s$ respectively, the change in the frequency of the A allele in a single generation follows equation

$$\Delta_p = p' - p = \frac{pqs(ph + q(1-h))}{\bar{w}},$$

where $q = 1 - p$ and $\bar{w} = 1 - 2pqhs - q^2s$ is the mean fitness of the population. When $0 < h < 1$, because Δ_p is always positive or negative (depending on the sign of s), the frequency of allele A will continuously increase or decrease. This model is called directional selection.

When $h < 0$, Δ_p can be positive or negative, depending on the allele frequency. If the heterozygous genotype has a greater fitness than either homozygous genotype, the allele will reach a stable allele frequency $\hat{p} = \frac{1-h}{1-2h}$ (solved from $\Delta_p = 0$) because $\Delta_p < 0$ when $p > \hat{p}$ and $\Delta_p > 0$ when $p < \hat{p}$. This situation is called *overdominance* or *heterozygote superiority*.

■ EXAMPLE 2.8

This example evolves three populations of 10,000 individuals for 200 generations, all with an initial frequency of 0.5 at the first locus. It uses a `MapSelector` to assign fitness to individuals according to their genotypes at locus `0`. Mutants in these populations are subject to purifying ($s = 0.1$, $h = 0.5$), positive ($s = -0.1$, $h = 0.2$) , and balancing ($s = 0.1$, $h = -0.5$) selection, respectively.

The dynamics of allele frequencies in these populations are calculated and outputed using operators `Stat` and `PyEval`. Because of different types of selection pressure, allele 0 gets fixed in the first population, gets lost in the second population, and reaches an equilibrium allele frequency $\hat{p} = 0.75$ in the third population.

The trajectory of allele frequencies in these three populations is also plotted using a `VarPlotter` operator (Figure 2.2). A number of parameters are passed to this operator to customize the look and feel of the figures. In particular, a parameter `lines_lty_rep=[1,2,3]` is used to pass a list of values to the `lines` function that draw allele frequencies for different replicates.

SOURCE CODE 2.8 Single-Locus Diallelic Selection Models

```
>>> import simuPOP as sim
>>> from simuPOP.plotter import VarPlotter
>>> pop = sim.Population(size=10000, loci=1, infoFields='fitness')
>>> simu = sim.Simulator(pop, rep=3)
>>> h = [0.5, 0.2, -0.5]
>>> s = [0.1, -0.1, 0.1]
>>> simu.evolve(
...     initOps=[
```

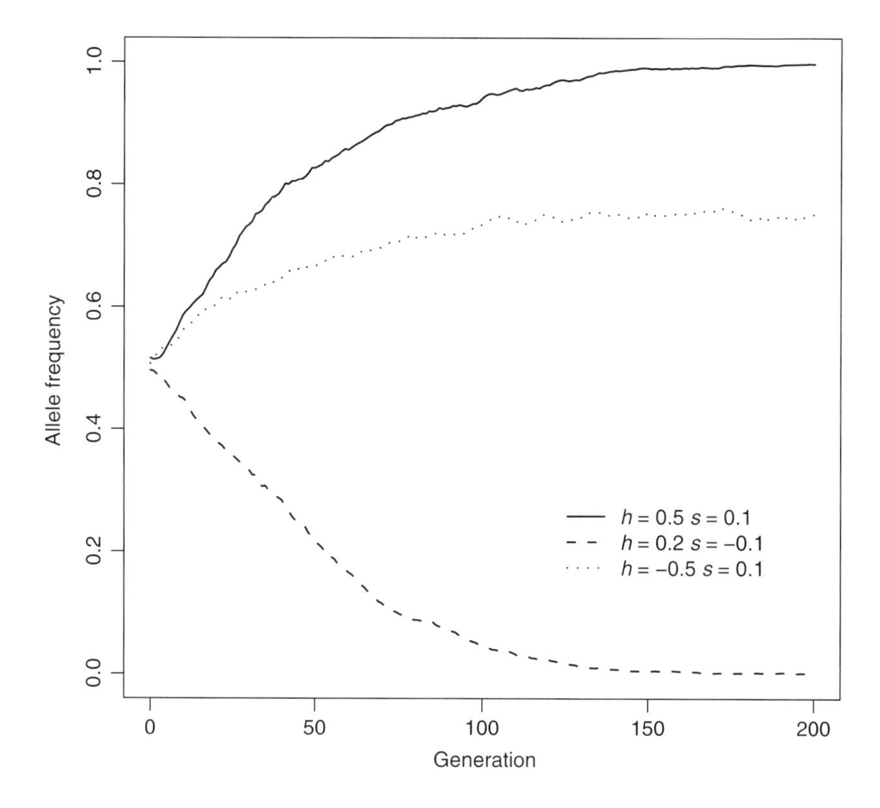

FIGURE 2.2 Trajectories of allele frequency for three populations under different selection pressures. Trajectories of the frequency of allele 0 for three populations that are subject to different selection pressures.

```
...         sim.InitSex(),
...         sim.InitGenotype(freq=(0.5, 0.5))
...     ],
...     preOps=[sim.MapSelector(loci=0,
...         fitness={(0,0):1, (0,1):1-h[x]*s[x], (1,1):1-s[x]}, reps=x)
...         for x in range(3)],
...     matingScheme=sim.RandomMating(),
...     postOps=[
...         sim.Stat(alleleFreq=0),
...         sim.PyEval(r'"%.3f\t" % alleleFreq[0][1]', step=50),
...         sim.PyOutput('\n', reps=-1, step=50),
...         VarPlotter('alleleFreq[0][0]', update=200,
...             legend=['h=%.1f s=%.1f' % (x,y) for x,y in zip(h,s)],
...             saveAs='Figures/selection.pdf',
...             lines_lty_rep=[1, 2, 3], lines_col='black',
...             lines_lwd=2, legend_x=120, legend_y=0.3,
...             plot_ylab='Allele Frequency', plot_ylim=[0,1]),
...     ],
...     gen=201
... )
```

```
0.484 0.504 0.493
0.173 0.785 0.334
0.058 0.955 0.264
0.010 0.995 0.251
0.002 0.999 0.248
(201, 201, 201)
```

2.6.2 Multilocus Selection Models

Multilocus selection models determine individual fitness from genotypes at multiple loci. In general, 3^n parameters will be needed to specify a diploid selection model with n loci. For example, a list of nine values are passed to operator `MaSelector` in Example 2.9 to simulate a symmetric viability model with recombination (Table 2.1) [3].

■ EXAMPLE 2.9

This example simulates a symmetric viability model with recombination. The population consists of 10,000 individuals, each with a chromosome with two linked loci. These two loci are initialized independently with an allele frequency of 0.5, which result in an equal frequency of 0.25 for all haplotypes *AB*, *Ab*, *aB*, and *ab*..

During evolution, a multiallelic selection operator `MaSelector` is used to assign fitness value to all individuals according to their genotype at two loci, using a symmetric viability model with $a = 1$, $b = 1.5$, $c = 2.5$, $d = 4$. This example uses two `Stat` operators to calculate the population mean fitness, frequency of all haplotypes, and the linkage disequilibrium between these two loci. Note that mean fitness has to be calculated before mating and after the `MaSelector` because the fitness values are set by the `MaSelector` to all parents.

Because $a < b$ and $c < d$, the fitness of a genotype increases with the number of heterozygous loci. Consequently, the frequencies of haplotypes Ab and aB increase during evolution that improves the mean fitness of the population and causes linkage disequilibrium between these two loci.

TABLE 2.1 A symmetric Viability Model of Natural Selection.

Fitness	BB	Bb	bb
AA	a	b	a
Aa	c	d	c
aa	a	b	a

SOURCE CODE 2.9 A Two-Locus Symmetric Viability Model of Natural Selection

```
>>> import simuPOP as sim
>>> pop = sim.Population(size=10000, loci=2, infoFields='fitness')
>>> a, b, c, d - 1, 1.5, 2.5, 4
>>> r = 0.02
>>> pop.evolve(
...     initOps=[
...         sim.InitSex(),
...         sim.InitGenotype(freq=(0.5, 0.5)),
...         sim.PyOutput('LD,   AB,   Ab,   aB,   ab,   avg fitness\n'),
...     ],
...     preOps=[
...         sim.MaSelector(loci=[0, 1], wildtype=0,
...             fitness=[a, b, a, c, d, c, a, b, a]),
...         sim.Stat(meanOfInfo='fitness'),
...     ],
...     matingScheme=sim.RandomMating(ops=sim.Recombinator(rates=r)),
...     postOps=[
...         sim.Stat(haploFreq=[0,1], LD=[0,1], step=20),
...         sim.PyEval(r"'%.3f %.2f %.2f %.2f %.2f %.2f\n' % (LD[0][1], "
...             "haploFreq[(0,1)][(0,0)], haploFreq[(0,1)][(0,1)],"
...             "haploFreq[(0,1)][(1,0)], haploFreq[(0,1)][(1,1)],"
...             "meanOfInfo['fitness'])", step=20),
...     ],
...     gen=100
... )
LD,   AB,  Ab,  aB,  ab,  avg fitness
0.001 0.25 0.25 0.26 0.25 2.25
0.022 0.23 0.27 0.27 0.23 2.26
0.090 0.16 0.34 0.34 0.16 2.27
0.178 0.07 0.43 0.43 0.07 2.36
0.205 0.05 0.45 0.46 0.04 2.43
100
```

2.7 GENEALOGY OF FORWARD-TIME SIMULATIONS

2.7.1 Genealogy of Haploid Simulations

Due to the stochastic nature of the parent selection process, it is likely that some parents will have one or more offspring, while some parents will have none. Assuming that there are N haploid parents (sequences) in the parental generation, let X be the number of distinct reproducing parents and k be the number of mating events, also the number of offspring, we have

$$\begin{aligned} &\Pr(X = p \mid k, N) \\ &= \Pr(X = p - 1 \mid k - 1, N)\,\frac{N - (p - 1)}{N} + \Pr(X = p \mid k - 1, N)\,\frac{p}{N} \end{aligned}$$

because the kth offspring has to inherit its genotype from a new parent if there are $i-1$ distinct parents after $k-1$ mating events, and it has also to inherit from an existing parent if there are already i distinct parents. Taking expectation at both sides and assuming $k \geq 1$ ($E(X \mid 0, N) = 0$), because there can be at most k distinct parents for k mating events, the expected number of reproducing parents after k mating events is

$$\begin{aligned} E(X \mid k, N) &= \sum_{p=1}^{k} p\Pr(X = p \mid k, N) \\ &= N\left(1 - \left(1 - N^{-1}\right)^{k}\right). \end{aligned}$$

It is easy to see that $\lim_{N\to\infty} \frac{E(X|N,N)}{N} = 1 - e^{-1} \sim 0.632$. That is to say, on average, only 63.2% of the parents will have the opportunity to pass their genotypes to the offspring generation in a haploid random mating scheme.

■ EXAMPLE 2.10

simuPOP is capable of recording one or more ancestral generations in a population during evolution. If the parents of all offspring are recorded, the complete pedigree information becomes available from which we can analyze how genotypes are transmitted from ancestors to the present population. This process involves the use of parameter `ancGen` to create populations that store specified number of ancestral generations and the use of operators `IdTagger` and `PedigreeTagger` to assign unique IDs to individuals and record IDs of offspring in their information fields. Refer to Section A.2.7 for details.

After a population is created, it can be converted to a `Pedigree` object using function `Population.asPedigree`. A `Pedigree` object is basically a `Population` object with additional capacity to refer to individuals by their IDs, and additional functions to analyze relationship between individuals. For example, if you would like to know the offspring of all individuals, you can call function

```
ped.locateRelatives(OFFSPRING, resultFields=offFields)    .
```

to locate all offspring of each individual and put their IDs in specified information fields `offFields`. Other relatives such as spouses (it is common to have multiple spouses when random mating is used), siblings (share at least one parent), and full siblings (share two parents) can be identified similarly, and more distant relationship can be derived from these relationships. In addition, this class also provides functions to identify parents or offspring of specified individuals across several generations.

Example 2.10 evolves a population of 1000 sequences forward in time for 1000 generations, using a haploid Wright–Fisher model. It uses a `IdTagger` to assign a unique ID to all individuals and a `parentTagger` with one information field `father_id` to record the parent of each offspring.

A function `Pedigree.identifyAncestors()` is used to identify all ancestors of individuals at the present generation. The next several lines remove all individuals except for these related ancestors. The population size of the remaining population is returned and printed. Because there are always less parents than their offspring, fewer and fewer ancestors contribute to the last generation if we look backward in time. It is not surprising that a small number of individuals in the starting population are ancestors of all individuals in the present population. This is essentially why coalescent simulations are much more efficient for simulating individual loci than the corresponding forward-time simulations.

SOURCE CODE 2.10 Number of Ancestors of a Haploid Simulation

```
>>> import simuPOP as sim
>>> pop = sim.Population(1000, ploidy=1, ancGen=-1,
...             infoFields=['ind_id', 'father_id'])
>>> pop.evolve(
...     initOps=[
...         sim.InitSex(),
...         sim.IdTagger(),
...     ],
...     matingScheme=sim.RandomSelection(
...         ops=[
...             sim.IdTagger(),
...             sim.PedigreeTagger(infoFields='father_id')
...         ],
...     ),
...     gen = 1000
... )
1000
>>> # a pedigree with only paternal information
>>> pop.asPedigree(motherField='')
>>> IDs = pop.identifyAncestors()
>>> allIDs = [ind.ind_id for ind in pop.allIndividuals()]
>>> removedIDs = list(set(allIDs) - set(IDs))
>>> pop.removeIndividuals(IDs=removedIDs)
>>> # number of ancestors...
>>> sizes = [pop.popSize(ancGen=x) for x in range(pop.ancestralGens())]
>>> print(sizes[0], sizes[100], sizes[500], sizes[999])
(1000, 18, 4, 3)
```

2.7.2 Genealogy of Diploid Simulations

In the diploid case, each mating event requires two parents chosen from their own respective sex groups, we have

$$E\left(X \mid k, N_m, N_f\right) = \left[N_m\left(1-\left(1-N_m^{-1}\right)^k\right) + N_f\left(1-\left(1-N_f^{-1}\right)^k\right)\right]$$

or $N\left(1-(1-2/N)^{k/c}\right)$ if we assume equal number of male and female parents, and that each mating event produces c offspring. When $c=2$, the ratio of expected number of parents to the number of offspring approaches $\lim_{N\to\infty}\frac{E(X|2N,N,N)}{2N} = 1-e^{-2} \sim 0.865$. This implies that on average 86.5% of parents are needed to produce an offspring population of the same size [4, 5]. More importantly, about 60% of parents are needed to produce an offspring population with half the size of the parental generation. The balance proportion is at approximately 80%, meaning on average 80% of parents are needed to produce 80% of offspring in the offspring generation. If we trace the number of ancestors who contribute their genotypes to the last generation backward in time, roughly 80% of all ancestors will be involved. Interestingly, even if we only need to simulate 5% of individuals of the last generation, more and more ancestors will contribute their genotypes to these 5% of individuals if we trace back in time, and eventually 80% of all ancestors will be involved.

■ EXAMPLE 2.11

We repeat the simulation performed in Example 2.10, this time for a diploid population using a `RandomMating` mating scheme. Instead of 1000 sequences, this example evolves a population of 1000 diploid individuals. Parentship is recorded during evolution using operators `IdTagger` and `PedigreeTagger`. After evolution, the pedigree object is trimmed to remove all ancestors who are not related to individuals in the present generation.

The result shows that there are 804 individuals in the starting population who are ancestors of the present population. Although these individuals might not contribute genetically to the present generation when the chromosome sequences are short, most of them will contribute their genotypes to the present population when chromosomes are long so that pieces of both maternal and paternal chromosomes will be passed to an offspring population because of genetic recombination.

SOURCE CODE 2.11 Number of Ancestors of a Diploid Simulation

```
>>> import simuPOP as sim
>>> pop = sim.Population(1000, ancGen=-1,
...     infoFields=['ind_id', 'father_id', 'mother_id'])
>>> pop.evolve(
...     initOps=[
...         sim.InitSex(),
...         sim.IdTagger(),
...     ],
...     matingScheme=sim.RandomMating(
...         ops=[
...             sim.IdTagger(),
...             sim.PedigreeTagger()
...         ],
...     ),
...     gen = 1000
... )
1000
>>> # a pedigree with only paternal information
>>> pop.asPedigree()
>>> IDs = pop.identifyAncestors()
>>> allIDs = [ind.ind_id for ind in pop.allIndividuals()]
>>> removedIDs = list(set(allIDs) - set(IDs))
>>> pop.removeIndividuals(IDs=removedIDs)
>>> # number of ancestors...
>>> sizes = [pop.popSize(ancGen=x) for x in range(pop.ancestralGens())]
>>> print(sizes[0], sizes[100], sizes[500], sizes[999])
(1000, 801, 786, 804)
```

This result has significant practical importance. If the simulated genome sequence is long enough so that almost all offspring chromosomes inherit part of their genotypes from both parental copies of parental chromosomes, 80% of all ancestor chromosomes will be involved in an ancestor recombination graph. This will be the case even when only a small sample is simulated using a coalescent approach. Only when there are very few recombination events during the evolutionary process can this graph be significantly reduced. This is essentially why coalescent simulations are no longer significantly more efficient than forward-time simulations in the simulation of long sequences.

REFERENCES

1. D. L. Hartl and A. G. Clark, *Principles of Population Genetics*, Sinauer Associates, Inc, 1997.
2. M. Kimura and T. Ohta, Stepwise mutation model and distribution of allelic frequencies in a finite population. *Proc Natl Acad Sci USA*, 75(6):2868–2872, 1978.

3. R.C. Lewontin and K. Ken-ichi, The evolutionary dynamics of complex polymorphisms. *Evolution*, 14:458–472, 1960.
4. C. Wiuf and J. Hein, On the number of ancestors to a DNA sequence. *Genetics*, 147(3):1459–1468, 1997.
5. B. Derrida, S. C. Manrubia, and D. H. Zanette, On the genealogy of a population of biparental individuals. *J Theor Biol*, 203(3):303–315, 2000.

CHAPTER 3

ASCERTAINMENT BIAS IN POPULATION GENETICS

3.1 INTRODUCTION

Ascertainment bias in population genetics is usually studied in two contexts. One of them is discovery of polymorphic loci and is best illustrated by invoking the example of SNPs (single-nucleotide polymorphisms). Most of the published data on SNP sampling frequencies are obtained in a two-step process, where the first step involves discovering chromosomal locations of a number of SNPs, and the second one involves DNA sequencing of a sample of n chromosomes restricted to locations discovered in the first step. The first step is called SNP discovery or ascertainment and is based on the number of chromosomes smaller than n. As demonstrated in a number of studies, taking into account the ascertainment scheme is a very important aspect of SNP data analysis. For example, Polanski and Kimmel [1] derived expressions for modeling the way in which ascertainment modified SNP sampling frequencies and distorted inferences concerning the rate. A more recent study [2] considers chip-based high-throughput genotyping, which has facilitated genome-wide studies of genetic diversity. Many studies have utilized these large data sets to make inferences

Forward-time Population Genetics Simulations: Methods, Implementation, and Applications,
Bo Peng, Marek Kimmel, and Christopher I. Amos.

about the demographic history of human populations using measures of genetic differentiation such as F_{ST} or principal component analysis. However, again, the single-nucleotide polymorphism chip data suffer from ascertainment biases caused by the SNP discovery process in which a small number of individuals from selected populations are used as discovery panels. Albrechtsen et al. [2] generate SNP genotyping data for individuals that previously have been subject to partial genome-wide Sanger sequencing and compare inferences based on genotyping data with inferences based on direct sequencing. They demonstrate that the ascertainment biases will distort measures of human diversity and possibly change conclusions drawn from these measures in some unexpected ways. They also show that details of the genotyping calling algorithms can have a surprisingly large effect on population genetic inferences. This type of ascertainment bias will be of importance in forthcoming genetic and genomic studies and the role of forward simulations will be rather serious.

However, in this chapter, we would like to show application of SimuPOP to ascertainment bias occurring in the interspecies or interpopulation studies. If any genetic measure of variability or diversity (such as heterozygosity) and its underlying cause (such as mutation rate) is studied in more than one species, a careful consideration of selection of the portions of the genome that are used as a basis for comparison is needed. Depending on from which species the polymorphisms are ascertained, the comparison of variability between the two species may be biased. We will consider a specific scenario, in which two extant species, such as humans and chimpanzees, are traced to a common ancestral species. Also, we will consider microsatellite loci, which can be modeled mathematically in a relatively simple way, so that the forward-time simulations can be compared with exact computations. Also, data are available in the literature to make the example interesting from the population genetics viewpoint.

In our example, ascertainment bias of interspecies (population) studies of microsatellite loci occurs when a locus is selected based on its large allele size in one species, in which it is first discovered (say, the cognate species 1). This bias is reflected in average allele length in any noncognate species 2 being smaller than that in species 1. This phenomenon was observed in various pairs of species, including comparisons of allele sizes in human and chimpanzee. Various mechanisms were proposed to explain the ascertainment bias. Here, we examine the simplest possible framework: A single-step asymmetric and unrestricted stepwise mutation model with genetic drift. The mathematical model is analyzed based on coalescence theory. The mechanism of ascertainment bias in this model is a tighter correlation of allele sizes within a cognate species 1 than of allele sizes in two different species 1 and 2. We present computations of the expected bias, given the mutation rate, population sizes of species 1 and 2, time of separation of species 1 and 2, and the age of the allele. In particular, using the coalescence theory, we show that when the past demographic histories of the cognate and noncognate taxa are different, the rate and directionality of mutations will impact the allele sizes in the two taxa differently than the simple effect of ascertainment bias.

Microsatellite polymorphisms, characterized by variations of copy numbers of short motifs of nucleotides, have become a common tool for gene mapping and evolutionary studies since they are abundantly found in genomes of a large number of organisms [3–6]. High mutation rates at these loci are the attractive feature of using the microsatellites as desired tools for molecular evolutionary studies, since consequences of accumulation of past mutation events are easily expected to be seen as differences of allele frequency distributions even in closely related taxa [7–9]. However, in cross-species comparisons of allele size distributions at microsatellite loci, some apparently discordant findings (namely, a systematic bias of average allele sizes in one species compared to another) led some investigators to argue that these repeat loci may not be the most efficient tools for interspecies studies [10, 11]. In general, for evolutionary studies, microsatellite loci identified in one species (or population) are studied in other species (or populations), making use of their comparative genome homology. Nevertheless, the process of detection (in the cognate taxon) and its use in a noncognate taxon may inherently impact the allele size distribution and other associated summary measures of genetic variation (such as heterozygosity, allele size variance, or number of segregating alleles). This discordance, called the ascertainment bias, is claimed to have been observed in sheep [12], swallows, cetaceans, ruminants, turtles, and birds [13]. However, Rubinsztein et al. [10] and Amos and Rubinsztein [14] explained such observations as intertaxa differences of rates and patterns of mutations at microsatellite loci.

The goal of the present investigation is to address this issue. Our approach is different from other attempts to study similar problems (see Ref. [15]), since we consider a general model of mutations (called the generalized stepwise mutation model (GSMM)) that is shown to be applicable for microsatellites [8, 16], on which we superimpose the effects of demographic differences of cognate and noncognate taxa, as both of these factors are known to jointly affect the features of polymorphisms at microsatellite loci in extant taxa [17]. In particular, using the coalescence theory, we show that when the past demographic histories of the cognate and non-cognate taxa are different, the rate and directionality of mutations will impact the allele sizes in the two taxa differently than the simple effect of ascertainment bias.

3.2 METHODS

3.2.1 Evolution of a DNA Repeat Locus

We consider a DNA repeat locus originated t units of time ago (backward time t), and observed at present (time 0). Adjective "backward" will be usually omitted. The chromosomes containing the locus belong to one of the two populations (labeled 1 and 2), which diverged t_0 time units before present (time t_0) from an ancestral population (labeled 0). The essentials are depicted in Figure 3.1. The ancestral population consists of $2N_0$ chromosomes, and populations 1 and 2 of $2N_1$ and $2N_2$ chromosomes, respectively. We assume the time-continuous Fisher–Wright–Moran model [17]. At the locus considered, alleles mutate according to the unrestricted generalized stepwise mutation model [8]. Specifically, the action of genetic drift and mutation can be represented by the following coalescence/mutation model:

1. Chromosomes 1 and 2, sampled at time 0 from populations 1 and 2, respectively, have a common ancestor T units of time before present (Figure 3.1). Random variable (rv) T has exponential distribution with parameter $1/(2N_0)$, shifted by t_0, that is,

$$\Pr[T > \tau] = \begin{cases} 1, & \tau \leq t_0, \\ \exp[-(\tau - t_0)/(2N_0)], & \tau > t_0. \end{cases} \tag{3.1}$$

   In other words, as long as the two chromosomes or their direct ancestors belong to different populations (i.e., for $\tau \leq t_0$, in backward

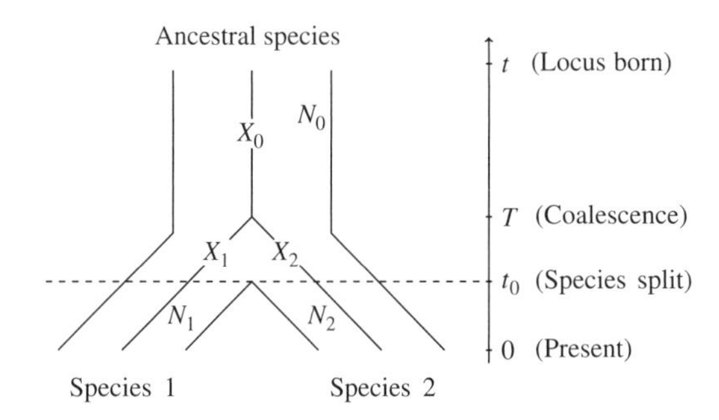

FIGURE 3.1 Demographic scenario employed in the mathematical model and SimuPOP simulations. N_0, N_1, and N_2 are effective sizes of the ancestral, cognate, and noncognate populations, respectively; X_0, X_1, and X_2 are increments of allele sizes due to mutations in the ancestral allele, in chromosome 1, and in chromosome 2, respectively.

time), they cannot coalesce. From the moment the populations converge (i.e., for $\tau > t_0$ in backward time), the distribution of the time to coalescence is exponential with parameter $1/(2N_0)$.

2. Chromosomes 1 and $1'$, sampled at time 0 from population 1, have a common ancestor T units of time before present, either in population 1, if $T \leq t_0$, or in the ancestral population 0, if $T > t_0$. Therefore, the random variable T has a more complex exponential distribution of the form

$$\Pr[T > \tau] = \begin{cases} \exp[-\tau/(2N_1)], & \tau \leq t_0, \\ \exp[-t_0/(2N_1) - (\tau - t_0)/(2N_0)], & \tau > t_0. \end{cases} \tag{3.2}$$

In other words, as long as the two chromosomes or their direct ancestors belong to population 1 (i.e., for $\tau \leq t_0$, in backward time), they coalesce with intensity $1/(2N_1)$. From the moment the species converge (i.e., for $\tau > t_0$ in backward time), the coalescence intensity is $1/(2N_0)$.

3. Initial size (number of repeats) at the locus at time (t) of the origin of the locus is equal to a constant. Choosing this constant equal to 0 is not a restrictive assumption. In our model, we assume that before time t, there are no mutation events.

4. Mutation epochs along the lines of descent occur according to a Poisson process with constant intensities ν_0, ν_1, and ν_2 in populations 0, 1, and 2, respectively. Each mutation event alters the allele size S by adding to it a random number of repeats U, that is,

$$S \to S + U.$$

U is an integer-valued random variable with probability-generating function (pgf)

$$\varphi_k(s) = E(s^U) = \sum_{i=-\infty}^{\infty} \Pr[U = i] s^i.$$

The pgf $\varphi_k(s)$ and, equivalently, the distribution of U are generally different in each population k $(k = 0, 1, 2)$. Consequently, the change of the allele size, during a time interval of length Δt spent in population k, is a compound Poisson random variable with pgf $\exp\{\nu \Delta t[\varphi_k(s) - 1]\}$. For the asymmetric single-step stepwise mutation model (SSMM), we have

$$\varphi_k(s) = b_k s + d_k/s, \tag{3.3}$$

where $b_k = \Pr[U = 1]$ and $d_k = \Pr[U = -1]$ are the respective probabilities of expansion and contraction of the allele in a single mutation epoch.

3.2.2 Conditional Distributions and Ascertainment Bias of Allele Sizes

The main purpose of this section is to use the principles of the coalescence theory (as reviewed by Tavaré [18]) to derive conditional expected allele size at a chromosome, given the allele size on another chromosome sampled either from a different or from the same population as the original chromosome. This information is crucial for obtaining theoretical estimates of the ascertainment bias in conjunction with other effects.

Chromosomes Sampled from Populations 1 and 2 We use notation as in Figure 3.1: X_0, X_1, and X_2 denote the incremental changes of allele sizes in the ancestral chromosome 0, and in chromosomes 1 and 2, respectively. Conditionally on T, X_0, X_1 and X_2 are independent random variables. Let us note that while chromosome 0 always lives in population 0, chromosomes 1 and 2 begin their lives in population 0 and then continue in populations 1 and 2. Let $Y_1 = X_0 + X_1$ and $Y_2 = X_0 + X_2$ denote the allele sizes at time 0 (present time) at chromosomes 1 and 2, respectively. We want to find the expected allele size at chromosome 2, jointly with the allele size at chromosome 1 being equal to i (conditional on $\{T = \tau\}$),

$$\begin{aligned} \mathrm{E}[Y_2; Y_1 = i | T = \tau] &= \sum_j \mathrm{E}[X_0 + X_2; X_0 = j; Y_1 = i - j | T = \tau] \\ &= \mathrm{E}[X_2 | T = \tau] \Pr[Y_1 = i | T = \tau] \\ &\quad + \sum_j j \Pr[X_0 = j | T = \tau] \Pr[X_1 = i - j | T = \tau]. \end{aligned} \tag{3.4}$$

If we translate the above into the language of generating functions, we obtain

$$\begin{aligned} \sum_i \mathrm{E}[Y_2; Y_1 = i | T = \tau] s^i &= \mathrm{E}[X_2 | T = \tau] f_{X_0|T=\tau}(s) f_{X_1|T=\tau}(s) \\ &\quad + s f'_{X_0|T=\tau}(s) f_{X_1|T=\tau}(s). \end{aligned} \tag{3.5}$$

Chromosomes Sampled from Population 1 Using the same reasoning, we obtain

$$\sum_i \mathrm{E}[Y_1'; Y_1 = i|T = \tau]s^i = \mathrm{E}[X_1'|T = \tau] f_{X_0|T=\tau}(s) f_{X_1|T=\tau}(s) + s f'_{X_0|T=\tau}(s) f_{X_1|T=\tau}(s). \quad (3.6)$$

Probability Generating Functions and Expectations of Incremental Changes of Allele Sizes Random variables X_0, X_1, and X_2 result from compounding the Poisson process [19] of mutations, with varying intensities ν_0, ν_1, and ν_2, by matching distributions of allele size changes with pgf values $\varphi_0(s)$, $\varphi_1(s)$, and $\varphi_2(s)$, respectively. The choice of intensity and pgf depends on the population in which the chromosomes reside during a given time interval. Without getting into detail, we obtain

$$f_{X_0|T=\tau}(s) = \begin{cases} \exp\{(t - t_0)\nu_0[\varphi_0(s) - 1] + (t_0 - \tau)\nu_1[\varphi_1(s) - 1]\}, & \tau \le t_0, \\ \exp\{(t - \tau)\nu_0[\varphi_0(s) - 1]\}, & t_0 < \tau \le t, \\ 1, & \tau > t, \end{cases} \quad (3.7)$$

$$f_{X_i|T=\tau}(s) = \begin{cases} \exp\{\tau\nu_i[\varphi_i(s) - 1]\}, & \tau \le t_0, \\ \exp\{(\tau - t_0)\nu_0[\varphi_0(s) - 1] + t_0\nu_i[\varphi_i(s) - 1]\}, & t_0 < \tau \le t, \\ \exp\{(t - t_0)\nu_0[\varphi_0(s) - 1] + t_0\nu_i[\varphi_i(s) - 1]\} & \tau > t, \end{cases} \quad (3.8)$$

for $i = 1, 2$. Also, $f_{X_1'|T=\tau}(s) \equiv f_{X_1|T=\tau}(s)$. The conditional expected values are obtained by differentiation of respective pgf values and setting $s = 1$.

Computational Expressions for E[Y_2; $Y_1 = i$] and E[Y_1'; $Y_1 = i$] In the single-step stepwise mutation model, the pgf values $\varphi_0(s)$, $\varphi_1(s)$, and $\varphi_2(s)$ have the form as in Equation 3.3. We note the expansion

$$e^{\nu t[bs+d/s-1]} = \sum_{i \in Z} \beta_i s^i = \sum_{i \in Z} e^{-\nu t} I_i(2\nu t\sqrt{bd}) \left(\frac{b}{d}\right)^{i/2} s^i \quad (3.9)$$

valid for $|s| = 1$, where $I_i = I_{-i}$ is the modified Bessel function of the first type, of integer order i [20]. Using this expansion, it is possible to represent the right-hand sides of Equations 3.5 and 3.6 as power series in variable s. Finally,

$$\mathrm{E}[Y_2; Y_1 = i] = \int_0^\infty \mathrm{E}[Y_2; Y_1 = i|T = \tau] f_T(\tau) d\tau,$$

$$\mathrm{E}[Y_1'; Y_1 = i] = \int_0^\infty \mathrm{E}[Y_1'; Y_1 = i | T = \tau] f_T(\tau) d\tau,$$

where $f_T(\tau)$ is the distribution density of the time to coalescence, based on Equations 3.1 and 3.2, respectively. A computational expression for $\Pr[Y_1 = i]$ can be similarly obtained.

Suppose that a DNA repeat locus discovered in a genome search of population 1 is retained for further study if it has a minimum number of x repeats of the motif, that is, if

$$Y_1 \geq x.$$

This criterion is also a substitute measure of this locus's variability, and hence of its polymorphism. The reason is that, irrespective of directionality of mutational changes, in the GSMM model the extremes of repeat count are strongly positively correlated with variance of repeat count and heterozygosity at the locus. This latter is a consequence of the random walk mechanism of mutations in this model (for a discussion and references, see Ref. [8]).

If the locus is retained and a sample of n individuals from the noncognate population 2 is typed for this locus, then the expected values of the mean repeat count in the sample is equal to

$$\mathrm{E}\left(\frac{1}{n}\sum_{i=1}^{n} Y_{2i} \,|Y_1 \geq x\right) = \mathrm{E}[Y_2|Y_1 \geq x] = \frac{\sum_{i\geq x} \mathrm{E}[Y_2; Y_1 = i]}{\sum_{i\geq x} \Pr[Y_1 = i]}.$$

If a sample of n individuals of the cognate population 1 is typed for this locus, then the expected values of the mean repeat count in the sample is equal to

$$\mathrm{E}\left(\frac{1}{n}\sum_{i=1}^{n} Y_{1i}' \,|Y_1 \geq x\right) = \mathrm{E}[Y_1'|Y_1 \geq x] = \frac{\sum_{i\geq x} \mathrm{E}[Y_1'; Y_1 = i]}{\sum_{i\geq x} \Pr[Y_1 = i]}.$$

The ascertainment bias of the mean allele size can be defined as

$$B = \mathrm{E}[Y_2|Y_1 \geq x] - \mathrm{E}[Y_1'|Y_1 \geq x]. \tag{3.10}$$

3.2.3 Simulation Method

Despite the complexity of the theory involved in the study of ascertainment bias, simulation of such a process is straightforward. Example 3.1 explains how to evolve a founder population for t generations and evolve again from generation $t - t_0$ if a random allele from the simulated population exceeds a specified threshold.

■ EXAMPLE 3.1

This example creates a founder population of size N. This diploid population has a locus with two alleles with an initial allele of 100. The founder population is evolved for $t - t_0$ generations before a copy of this population is expanded to a size of N_1 and is evolved for another t_0 generations. The original population will also evolve t_0 generations if a random allele from this population has more tandem repeats than a specified threshold. The difference between the mean number of tandem repeats in these two populations will be returned as a measure of ascertainment bias. During this evolutionary process, a step-wise mutation model with different mutation rates and increasing probabilities can be specified for the evolution of the founder and two split populations.

SOURCE CODE 3.1 Script to Simulate the Evolution of Microsatellite Marker Using a Scaling Technique

```
def simuAscerBias(t, t0, N, N1, N2, v0, v1, v2, incProb, thresh):
    '''This function evolves a founder population of N individuals for
    t-t0 generations. A copy of this population is expanded to a size of
    N1 and continued to evolve for t-t0 generations. If a randomly chosen
    allele from the this population has more tandem repeats than a
    specified threshold, the same copy of population (at generation
    t-t0) is evolved similarly with a population size of N1. Ascertainment
    bias is returned as the length difference between the first and
    second populations.'''
    while True:
        # Evolve an ancestral population with size N0
        pop = sim.Population(size=N, loci=1)
        pop.evolve(
            initOps=[
                sim.InitSex(),
                sim.InitGenotype(genotype=[100])
            ],
            preOps=sim.StepwiseMutator(rates=v0, incProb=incProb, loci=0),
            matingScheme=sim.RandomMating(),
            gen=t-t0
        )
        # make a copy of pop
        pop1 = pop.clone()
        # Evolve for another t0 generations with population size N1
        pop1.evolve(
            preOps=sim.StepwiseMutator(rates=v1, incProb=incProb, loci=0),
            matingScheme=sim.RandomMating(subPopSize=N1),
            gen=t0
        )
```

```
        # draw a random allele from pop1
        ind1 = pop1.individual(randint(0, N1-1))
        # if the allele length is greater than the threshold,
        if ind1.allele(randint(0,1)) > thresh:
            # Evolve pop2 for t0 generations with population size N2
            pop.evolve(
                preOps=sim.StepwiseMutator(rates=v2, incProb=incProb, loci=0),
                matingScheme=sim.RandomMating(subPopSize=N2),
                gen=t0
            )
            return (sum(pop1.genotype()) - sum(pop.genotype()))/(2.*N1)
```

Direct execution of simulations described in Example 3.1 is however infeasible. Because populations are large, it is time-consuming to evolve them for tens of thousands of generations. What makes things worse is that the probability that a random allele exceeds a specified threshold can be low, meaning many attempts may be needed to obtain a successful measure of ascertainment bias.

This problem can be addressed through the use of a scaling technique [21]. Compared to a regular simulation that evolves a population of size N for t generations, a scaled simulation with a scaling factor λ evolves a smaller population of size N/λ for t/λ generations with magnified (multiplied by λ) mutation, recombination, and selection forces. This method could be justified by a diffusion approximation to the standard Wright–Fisher process [21, 22]; however, because the diffusion approximation only applies to weak genetic forces in the evolution of haploid sequences, it cannot be used when nonadditive diploid or strong genetic forces are used. Simulations for this chapter are performed with scaling factor of 100, where populations with sizes $N_i/100$ are evolved for $t_i/100$ generations, under mutation models with mutation rates $100\nu_i$. Running the simulations with different scaling factors yields identical results for $\lambda < 100$.

3.3 RESULTS

3.3.1 Summary of Modeling Results

The purpose of our modeling is to determine in what circumstances the presence or absence of differences, observed in sizes of alleles at loci discovered in a cognate species (population 1) and then typed in a noncognate species (population 2), can be attributed to ascertainment bias or alternatively to differential effects of genetic drift or mutation rate and pattern.

Before we present numerical result, let us review the intuitions concerning these effects. These intuitions are in a major part valid independent of a particular model of mutations:

1. Ascertainment bias *per se* results from a stronger correlation between allele states of chromosomes in cognate population 1 compared to the correlation between allele states in cognate population 1 versus noncognate population 2.
2. Genetic drift can reduce the effects of ascertainment bias. Indeed, if the cognate population 1 is much larger than the noncognate population 2, then the coalescence process within population 1 has the star-like structure characterized by reduced dependence of allele states [23]. Therefore, the difference in correlations of allele states of chromosomes in cognate population 1 compared to correlations of allele states in cognate population 1 versus noncognate population 2 will be reduced. Note that the size of the noncognate population 2 will not influence the difference of expected allele sizes, but it may influence other indices of polymorphisms.
3. Mutation rate and pattern, different in populations 1 and 2, obviously can influence the differences in allele sizes between chromosomes in different populations.

Figures 3.2 and 3.3 depict a series of modeling studies of B, the combined effect of ascertainment bias, genetic drift, and differential mutation rate on the mean repeat count based on SimuPOP model, compared to those obtained using Equation 3.10. Basic parameter values approximate the evolutionary dynamics of dinucleotides in humans and chimpanzees: Time from divergence of species $t_0 = 5 \times 10^6$ years $= 2 \times 10^5$ generations for Figure 3.2 and $t_0 = 6.25 \times 10^6$ years $= 2.5 \times 10^5$ generations for Figure 3.3, the age of the repeat locus in the range $t = 5 \times 10^5$ generations, mutation rate $\nu = 1 \times 10^{-4}$ per generation, and probability of increase of allele size in a single mutation event, $b = 0.55$. Effective size of the current human population is $2N = 4 \times 10^5$ individuals.

Figure 3.2 compares simulation results with theoretical estimates from Equation 3.10. It depicts the values of B for the basic parameter values $b_0 = b_1 = b_2 = b$ and $\nu_0 = \nu_1 = \nu = 0.0001$, with the effective sizes of all populations concurrently varying from 2×10^4 to 4×10^5 individuals and with mutation rates ν_1 (3.2.a) or ν_2 (3.2.b) varying from ν to 5ν. Figures 3.2 a and b makes it explicit that the combined effect of ascertainment bias,

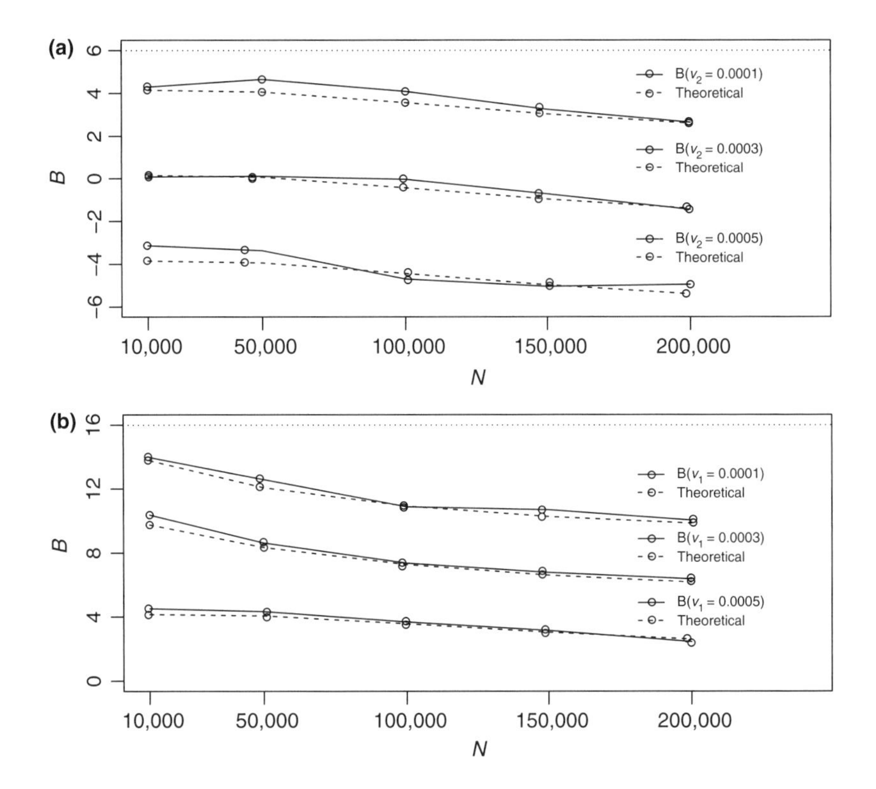

FIGURE 3.2 Comparison of SimuPOP simulations with mathematical expressions from Equation 3.10. (a) Values of B for the basic parameter values $b_0 = b_1 = b_2 = b$ and $\nu_0 = \nu_1 = \nu = 0.0001$, with the effective sizes of all populations concurrently varying from 2×10^4 to 4×10^5 individuals and with mutation rates ν_2 varying from ν to 5ν. (b) Values of B for the basic parameter values $b_0 = b_1 = b_2 = b$ and $\nu_0 = \nu_2 = \nu = 0.0001$, with the effective sizes of all populations concurrently varying from 2×10^4 to 4×10^5 individuals and with mutation rates ν_1 varying from ν to 5ν.

genetic drift, and differential mutation rate on the mean repeat count can result in a range of B values from positive to negative ones.

Figure 3.3a depicts the values of B for the basic parameter values $b_0 = b_1 = b_2 = b$, with the effective sizes of the ancestral and noncognate populations kept at a much lower level of $2N_0 = 2N_2 = 2 \times 10^4$ individuals, with the size of the cognate population kept at the basic level $2N_1 = 2N = 4 \times 10^5$. The lower effective population sizes $2N_0$ and $2N_2$ can be regarded as reflecting the lower population counts of common ancestors of humans and chimpanzees and of their chimpanzee

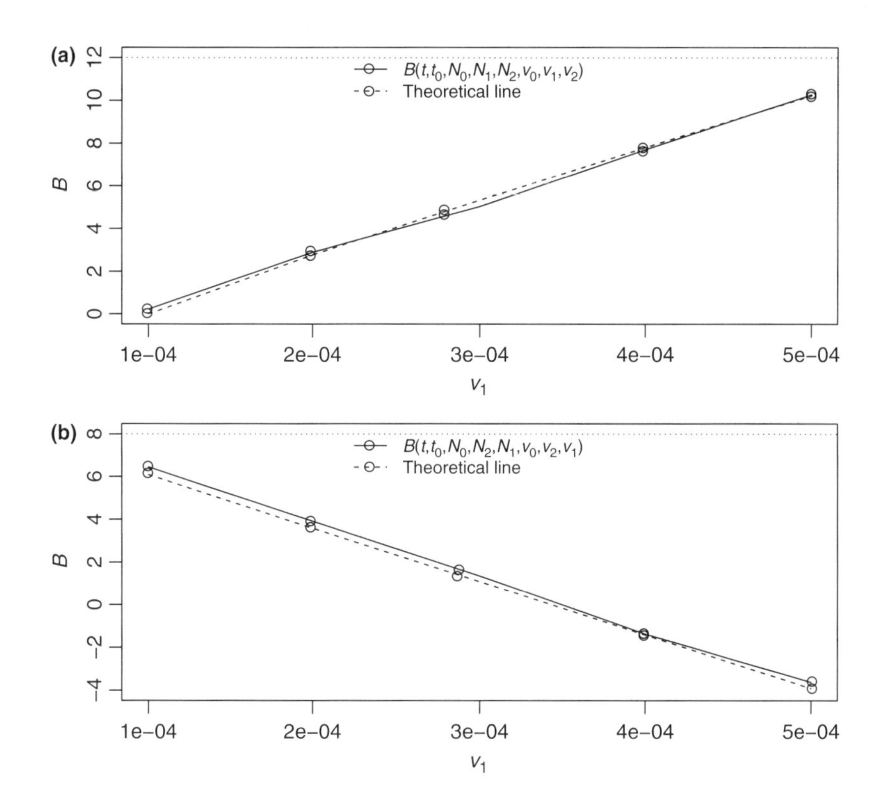

FIGURE 3.3 Comparison of SimuPOP simulations with mathematical expressions from Equation 3.10. (a) Values of B for the basic parameter values $b_0 = b_1 = b_2 = b$, with the effective sizes of the ancestral and noncognate populations kept at the level of $2N_0 = 2N_2 = 2 \times 10^4$ individuals, with the size of the cognate population kept at the basic level $2N_1 = 2N = 4 \times 10^5$. The mutation rate of the cognate (larger) population is varied from $\nu_1 = 1 \times 10^{-4}$ to 5×10^{-4} per generation, while the mutation rates of the ancestral and the noncognate (smaller) population are set at $\nu_0 = \nu_2 = 1.5 \times 10^{-4}$ per generation. (b) A reverse configuration: The mutation rate of the noncognate (larger) population is varied from $\nu_1 = 1 \times 10^{-4}$ to 5×10^{-4} per generation, while the mutation rates of the ancestral and the cognate (smaller) population are set at $\nu_0 = \nu_2 = 1 \times 10^{-4}$ per generation.

descendants. The mutation rate of the cognate (larger) population is varied from $\nu_1 = 1 \times 10^{-4}$ to 5×10^{-4} per generation, while the mutation rates of the ancestral and the noncognate (smaller) population are set at $\nu_0 = \nu_2 = 1.5 \times 10^{-4}$ per generation. At $\nu_1 \approx 3 \times 10^{-4}$, the value of B is approximately equal to 5, the difference of mean allele size observed when human loci are typed in chimpanzee, as described in the following section.

Figure 3.3b depicts the reverse configuration. Now, the effective sizes of the ancestral and cognate populations are kept at the level of $2N_0 = 2N_2 = 2 \times 10^4$ individuals, with the size of the noncognate population kept at the basic level $2N_1 = 2N = 4 \times 10^5$. The mutation rate of the noncognate (larger) population is varied from $\nu_1 = 1 \times 10^{-4}$ to 5×10^{-4} per generation, while the mutation rates of the ancestral and the cognate (smaller) population are set at $\nu_0 = \nu_2 = 1.5 \times 10^{-4}$ per generation. At $\nu_1 \approx 3 \times 10^{-4}$, the value of B is approximately equal to -1, the difference of mean allele size observed when chimpanzee loci are typed in human, as described in the following section.

3.3.2 Comparisons of Empirical Statistics Derived from Human and Chimpanzee Microsatellite Data

As an illustration of our model we use, without claiming generality, the well-known data published by Cooper et al [24]. These authors previously examined 40 human microsatellite markers and their homologues in a panel of nonhuman primates and showed that human loci tend to be longer, a trend that was apparent in several other studies. Taken at face value, these data indicate that since their most recent common ancestor, more microsatellite expansion mutations have occurred in the lineage leading to humans compared to the lineage leading to chimpanzees. Based on this, they suggested that this provided evidence that microsatellites tended to expand with time and were doing so more rapidly in humans. However, an alternative explanation is that the length differences are due to ascertainment bias arising from the selection of longer than average human loci as markers. Cooper et al [24] presented the necessary reciprocal experiment showing that human microsatellites tend to be longer than their chimpanzee homologues, regardless of the species from which the loci were cloned. The data comprised 38 chimpanzee-derived CA-repeat microsatellites that were amplified in a panel of six chimpanzees and six humans.

For loci that are polymorphic in both species, the difference between the reciprocal comparisons can be used to estimate the size of the ascertainment bias affecting human–chimpanzee comparisons. For dinucleotide repeat loci cloned and characterized in humans ($n = 22$), human loci were an average of 5.18 repeat units longer than in chimpanzees, while dinucleotide repeats cloned from chimpanzees ($n = 25$) were on average 1.23 repeat units longer in humans. As noticed in the previous section, these data are reproduced by our model assuming specific values of effective population sizes, mutation rates, and times to separation of species.

3.4 DISCUSSION AND CONCLUSIONS

Computations presented in this chapter demonstrate that the scaled forward simulations using SimuPOP closely match the analytical solution of the evolutionary model of Section 3.2.1. Let us note that the ability to perform the mathematical derivations of Section 3.2.2 depends on the simplicity of the assumed microsatellite discovery criterion $Y_1 \geq x$. If this criterion is replaced by a condition on heterozygosity or variance, the theoretical derivations become practically impossible. On the other hand, it is easy to use any other microsatellite discovery criterion in SimuPOP simulations.

Data of Cooper et al [24] analyzed by us show that when the human-derived dinucleotide repeat loci were typed in chimpanzee, they show a trend toward smaller mean allele sizes in the chimpanzee compared to that in human populations. These and other data also suggest that the same holds for other measures of within-population variation (i.e., the chimpanzees showing lower heterozygosity and allele size variance compared to humans). The theoretical model, discussed in Section 3.2.1, shows that these observations are in agreement with ascertainment bias, caused by a selective choice of the human-specific loci. In the reciprocal experiment, the chimpanzee-derived dinucleotides, typed in humans populations, also show a trend toward smaller mean allele sizes in the chimpanzee compared to that in human populations.

These observations imply that the ascertainment bias is a factor of an appreciable order of magnitude in interpreting interpopulation genetic variation at microsatellite loci, when the loci are selectively chosen for polymorphism in one of the populations being compared. However, its effect is confounded by other differences in evolutionary dynamics between the cognate and noncognate populations as well as interpopulation differences of rates of mutations at the loci. For example, under the assumption that dinucleotide loci evolving under a generalized stepwise mutation model [8] show that with mutation rate and population size being the same in cognate and noncognate populations, the mean allele size in the cognate population may be several repeat units larger. The bias decreases as population size (and mutation rate) increases (Figure 3.2a and b). However, our numerical computations show that differences of population size as well as mutation rate may enhance or decrease this ascertainment bias, depending upon the pattern of difference. As shown in Figure 3.2a, increased mutation rate in the noncognate population can reduce, or even reverse, the ascertainment bias. Likewise, increased mutation rate in the cognate population may amplify the bias (Figure 3.2b). Numerical evaluations of the model suggest that the primary reason of ascertainment bias is the tighter correlation of allele sizes within the cognate population. Thus, intuitively it is clear that population size differences between cognate and non-cognate populations may also erase or amplify ascertainment bias. If the cognate population is of larger size or is growing more rapidly than the noncognate one, a reduced bias is expected.

The differences of patterns of ascertainment biases seen at the human-specific versus chimpanzee-specific dinucleotide loci can be explained in terms of the above theoretical predictions, if the mutation rate is higher for humans as explained in Section 3.2.1. The observed pattern (namely, the ascertainment bias is of a lower magnitude for the chimpanzee-specific loci) is also consistent with effective population size in chimpanzees being smaller than that of human populations. In this sense, our observation and theoretical predictions are consistent with the assertion of Rubinsztein et al [10], although an expansion bias of mutations is not necessary to explain the observed differences in humans and chimpanzees.

Our theory and data can also be used to explain the apparently discordant conclusions reached by other investigators examining this issue. For example, Ellegren et al. [13] observed smaller allele sizes in noncognate species compared to that in cognates of birds, which could be predominantly due to ascertainment bias alone. Crawford et al. [11], in contrast, found longer median allele sizes in sheep (compared to that in cattle), regardless of the origin of the microsatellites. This may be the case where the ascertainment bias effect is erased (or even reversed) due to mutation rate and effective population size differences in sheep and cattle. Cooper et al. [24] also published reciprocal studies on ascertainment bias, analogous to our data of the present paper. Their observations are not at variance with our findings, as they also observed some chimpanzee-specific loci being significantly longer in humans.

There had been some discussions with regard to the dependence of inter-population allele size differences on the absolute repeat lengths of alleles [13, 14]. It is true that for microsatellites there is a general tendency of increased level of polymorphism at loci harboring larger alleles [25]. Our theory shows that loci exhibiting larger degree of polymorphism will be subject to lesser bias of ascertainment (due to looser correlation of allele sizes in the cognate population). Hence, appropriate adjustment of inter-locus differences of polymorphism as well as allele sizes should be made in addressing the importance of ascertainment bias. Our approach to analyzing of "centered" statistics at least partially circumvents this interlocus component of variability.

In summary, we conclude that ascertainment bias should be an important consideration for interpretation of interpopulation differences of genetic variation at microsatellite loci, but this bias can be reduced or even reversed when the past demographic histories of cognate and noncognate populations are drastically different. In addition, mutation rate differences in populations (either due to their reproductive behavior or differences of cell division during oogenesis and spermatogenesis) can also mimic ascertainment bias.

REFERENCES

1. A. Polanski and M. Kimmel, New explicit expressions for relative frequencies of single-nucleotide polymorphisms with application to statistical inference on population growth. *Genetics*, 165(1):427–436, 2003.
2. A. Albrechtsen, F. C. Nielsen, and R. Nielsen, Ascertainment biases in SNP chips affect measures of population divergence. *Mol Biol Evol*, 27(11):2534–2547, 2010.
3. S. D. Pena, P. C. Santos, M. C. Campos, and A. M. Macedo, Paternity testing with the F10 multilocus DNA fingerprinting probe. *EXS*, 67:237–247, 1993.
4. R. Deka, M. D. Shriver, L. M. Yu, L. Jin, C. E. Aston, R. Chakraborty, and R. E. Ferrell, Conservation of human chromosome 13 polymorphic microsatellite (CA)n repeats in chimpanzees. *Genomics*, 22(1):226–230, 1994.
5. A. M. Bowcock, A. Ruiz-Linares, J. Tomfohrde, E. Minch, J. R. Kidd, and L. L. Cavalli-Sforza, High resolution of human evolutionary trees with polymorphic microsatellites. *Nature*, 368(6470):455–457, 1994.
6. C. R. Primmer and H. Ellegren, Patterns of molecular evolution in avian microsatellites. *Mol Biol Evol*, 15(8):997–1008, 1998.
7. J. L. Weber and C. Wong, Mutation of human short tandem repeats. *Hum Mol Genet*, 2(8):1123–1128, 1993.
8. M. Kimmel and R. Chakraborty, Measures of variation at DNA repeat loci under a general stepwise mutation model. *Theor Popul Biol*, 50(3):345–367, 1996.
9. R. Chakraborty, M. Kimmel, D. N. Stivers, L. J. Davison, and R. Deka, Relative mutation rates at di-, tri-, and tetranucleotide microsatellite loci. *Proc Natl Acad Sci USA*, 94(3):1041–1046, 1997.
10. D. C. Rubinsztein, J. Leggo, and W. Amos, Microsatellites evolve more rapidly in humans than in chimpanzees. *Genomics*, 30(3):610–612, 1995.
11. A. M. Crawford, S. M. Kappes, K. A. Paterson, M. J. deGotari, K. G. Dodds, B. A. Freking, R. T. Stone, and C. W. Beattie, Microsatellite evolution: testing the ascertainment bias hypothesis. *J Mol Evol*, 46(2):256–260, 1998.
12. S. H. Forbes, J. T. Hogg, F. C. Buchanan, A. M. Crawford, and F. W. Allendorf, Microsatellite evolution in congeneric mammals: domestic and bighorn sheep. *Mol Biol Evol*, 12(6):1106–1113, 1995.

13. H. Ellegren, C. R. Primmer, and B. C. Sheldon, Microsatellite "evolution": directionality or bias? *Nat Genet*, 11(4):360–362, 1995.
14. W. Amos and D. C. Rubinsztein, Microsatellites are subject to directional evolution. *Nat Genet*, 12(1):13–14, 1996.
15. A. R. Rogers and L. B. Jorde, Ascertainment bias in estimates of average heterozygosity. *Am J Hum Genet*, 58(5):1033–1041, 1996.
16. M. Kimmel, R. Chakraborty, D. N. Stivers, and R. Deka, Dynamics of repeat polymorphisms under a forward–backward mutation model: within- and between-population variability at microsatellite loci. *Genetics*, 143(1):549–555, 1996.
17. M. Kimmel, R. Chakraborty, J. P. King, M. Bamshad, W. S. Watkins, and L. B. Jorde, Signatures of population expansion in microsatellite repeat data. *Genetics*, 148(4):1921–1930, 1998.
18. S. Tavaré, Line-of-descent and genealogical processes, and their applications in population genetics models. *Theor Popul Biol*, 26(2):119–164, 1984.
19. J. F. C. Kingman, *Poisson Processes*, Oxford University Press, Oxford, 1993.
20. M. Abramowitz and I. Stegun, *Handbook of Mathematical Functions, with Formulas, Graphs, and Mathematical Tables*, U.S. Government Printing Office, 1972.
21. C. J. Hoggart, M. Chadeau-Hyam, T. G. Clark, R. Lampariello, J. C. Whittaker, M. De Iorio, and D. J. Balding, Sequence-level population simulations over large genomic regions. *Genetics*, 177(3):1725–1731, 2007.
22. Warren J. Ewens, Mathematical Population Genetics, Springer, 2004.
23. F. Tajima, The effect of change in population size on DNA polymorphism. *Genetics*, 123(3):597–601, 1989.
24. G. Cooper, D. C. Rubinsztein, and W. Amos, Ascertainment bias cannot entirely account for human microsatellites being longer than their chimpanzee homologues. *Hum Mol Genet*, 7(9):1425–1429, 1998.
25. J. L. Weber, Informativeness of human (DC-DA)n.(DG-DT)n polymorphisms. *Genomics*, 7(4):524–530, 1990.

CHAPTER 4

OBSERVING PROPERTIES OF EVOLVING POPULATIONS

Coalescent-based methods only simulate individuals in a coalescent tree, namely, people who are genetically related to the sample that is simulated. Because only a small fraction of ancestors is simulated, it is not possible to gather population properties of the ancestral populations during simulation. On the other hand, because these methods are based on standard Wright–Fisher models, the properties of the ancestral populations can be derived theoretically, so there is less demand to study them through simulations. In contrast, forward-time simulation methods evolve populations generation by generation, so it is easy to observe population properties during an evolutionary process. This is especially useful for the study of evolutionary processes that cannot be completely characterized by a Wright–Fisher model. Because these processes are usually mathematically intractable, forward-time simulations often remain the only effective method to study the properties of these processes.

In this chapter, we use forward-time population genetics simulations to follow the evolution of human genetics diseases and investigate the impact of various genetic and demographic factors on the allelic spectra of human diseases, based on a model proposed by Reich and Lander [1]. Because the goal of this chapter is to demonstrate how to observe the properties of

Forward-time Population Genetics Simulations: Methods, Implementation, and Applications,
Bo Peng, Marek Kimmel, and Christopher I. Amos.

evolving populations, we present simulation results only briefly. Interested readers should refer to Refs [2, 3] and other references for more details on this topic.

4.1 INTRODUCTION

4.1.1 Allelic Spectra of Complex Human Diseases

Allelic spectrum of a gene refers to the number and frequency of alleles at this gene. A spectrum is called *simple* if it contains a few common alleles that are carried by a majority of the population, and *diverse* if it contains many rare alleles (Figure 4.1). For example, three mutations in the BRCA1 and BRCA2 genes account for approximately 90% of the BRCA1 and BRCA2 mutations identified in Ashkenazi Jews. In contrast, a diverse spectrum can have many rare alleles, which is a relevant way to characterize BRCA1 and BRCA2 in non-Ashkenazi European populations [4]. Understanding why allelic spectra of the same gene would differ in different human populations is one of the goals of this chapter.

The nature of the allelic spectra of genes involved in a genetic disease is crucial for the success of mapping these genes using association or linkage disequilibrium (LD) methods because the statistical power of these methods will be greatly reduced if a gene contains many rare alleles (for impact of allelic heterogeneity, see Section 1.3.3). Two major hypotheses have been proposed to describe the genetic structure of complex human diseases:

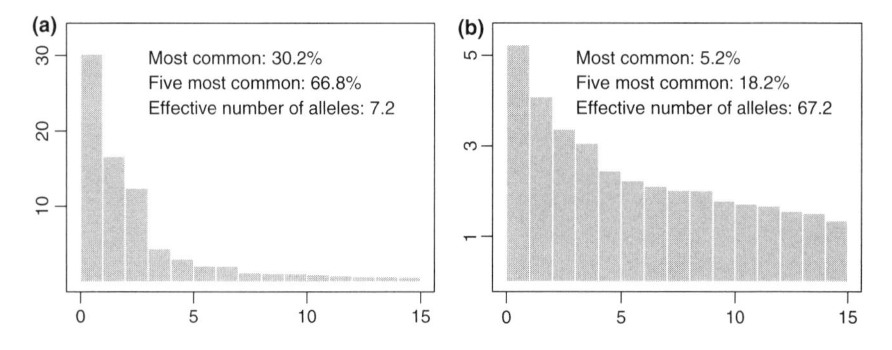

FIGURE 4.1 Examples of simple and diverse allelic spectra. Examples of simple (a) and diverse (b) allelic spectra. Fifteen most frequent alleles are plotted for both spectra, ordered by allele frequencies (y-axis). Effective number of alleles of both spectra are estimated using Equation 4.1 with $\sum f_i = 1$.

while the common disease–common variant (CDCV) hypothesis proposes that common diseases are usually caused by one or a few common disease susceptibility alleles at each disease-predisposing locus [5], the common disease–rare variants (CDRV) hypothesis proposes that common diseases might be caused by many rarc alleles, each having a relatively large impact on the disease [6, 7]. simuPOP-based mapping methods can be used to provide data supporting one hypothesis or the other.

4.1.2 An Evolutionary Model of Effective Number of Disease Alleles

Allelic diversity of a locus can be summarized by its *effective number of alleles* [8], which is reciprocal of the *expected allelic identity*, namely, the probability that two randomly chosen alleles are identical. Because two randomly chosen alleles are likely to be different if there are many rare alleles, a diverse spectrum will have a high effective number of alleles.

Genetic diversity of disease alleles can be summarized by a similar concept called *effective number of disease alleles*, which is defined as the reciprocal of the expected allelic identity among disease alleles. Assuming that allele 0 is the wild-type allele and all others are disease alleles, the effective number of disease alleles at a locus can expressed as

$$n_{\mathrm{e}} = \frac{1}{\phi_{\mathrm{dis}}} = \frac{(1 - f_0)^2}{\sum_{i>0} f_i^2}, \tag{4.1}$$

where ϕ_{dis} is the expected allelic identity among disease alleles and f_i is the allele frequency of allele i, $i = 0, 1, \ldots$. If there are k alleles at a locus (including a wild-type allele 0), n_{e} reaches its maximum value of $k - 1$ when all disease alleles have the same frequency, regardless of the frequency of the wild-type allele. If a population is in mutation–selection equilibrium, the equilibrium effective number of disease alleles at a disease susceptibility locus is

$$n_{\mathrm{e}} = 1 + 4N\mu\,(1 - f_{\mathrm{e}}),$$

where N is the effective population size, μ is the mutation rate under an infinite allele mutation model, and f_{e} is the equilibrium total disease allele frequency [9].

Reich and Lander [1] models the evolution of the effective number of disease alleles (n_{e}) using an instant population growth model. Assuming that the total disease allele frequency $1 - f_0 = \sum_{i>0} f_i$ is not far from its equilibrium value f_e during evolution, the estimated n_e, under the infinite allele model, is

$$\hat{n}_e = 1 + 4N\mu f_0, \tag{4.2}$$

where N is the effective population size and μ is the mutation rate. The equilibrium total disease allele frequency f_e is determined by the nature of disease. For example, the equilibrium value of the total disease allele frequency of recessive diseases can be approximated by

$$f_e = \sqrt{\mu/s}, \tag{4.3}$$

where s is the selection coefficient, provided that $s \gg \mu$.

For a population that has expanded instantly from size N_0 to N_1, the proportion of alleles derived from preexpansion will decay exponentially with the rate $\left(\frac{1-f_e}{f_e}\right)\mu$. The effective number of disease alleles will increase with expectation:

$$n_e(t) = \left(n_{e1}^{-1} + \left(n_{e0}^{-1} - n_{e1}^{-1}\right)\exp\left(-\frac{n_{e1}}{2N_1 f_e}t\right)\right)^{-1}, \tag{4.4}$$

where $n_{e0} = 1 + 4N_0\mu_s(1 - f_e)$ and $n_{e1} = 1 + 4N_1\mu_s(1 - f_e)$ [1]. Although the allelic spectra of both rare and common diseases are similar in equilibrium states, the rates at which these are approached differ greatly because the rate of reaching equilibrium is determined by the total disease allele frequency (f_e) within the exponential term of Equation 4.4. If the population has not reached a mutation selection equilibrium, common diseases at the beginning of the population expansion are likely to have simpler spectra than the rare diseases. This point will be further illustrated with simulations in the later sections.

4.1.3 Simulation of the Evolution of n_e

Mathematical analysis of the evolution of n_e led to the conclusion that CDCV hypothesis holds and that the phenomenon is caused by transient effects of demography (population expansion) [1]. Although in the long run when human population reaches mutation, selection, and drift equilibrium, all diseases will have diverse spectra, these authors argue that common diseases tend to have simpler spectra because they diversify their spectra more slowly than the rare diseases.

However, the evolution of human populations is much more complex than what is assumed in this model and it is unclear whether or not Reich and

Lander's conclusions [1] still hold when alternative or additional genetic features are incorporated into the model. Because mathematical analyses of the extended models are prohibitively difficult, forward-time simulations are used to study the evolution of n_e under more realistic models. Among the features considered are more complex demographics, finite allele mutation models, population structure and migration, and population size.

4.2 SIMULATION OF THE EVOLUTION OF ALLELE SPECTRA

4.2.1 Demographic Models

Reich and Lander [1] used a simple instant population growth model. Under the model assumption, a population with an initial population size N_0 would evolve G_0 generations, instantly expand its size to N_1, and continue to evolve for another G_1 generation. This model is definitely unrealistic, but it provides a mathematically tractable case that provides theoretical results for an extreme case to which other models can be compared. Because an exponential population expansion model is more realistic for most populations, we will simulate both instantaneous and exponential population expansion models. Example 4.1 demonstrates how to implement a demographic model in simuPOP.

■ EXAMPLE 4.1

The easiest method of implementing a demographic model is to define a demographic function, which is a user-defined Python function that accepts a generation number and a parental population, and returns a list of subpopulation sizes of the corresponding offspring population. A single number can be returned if there is no population structure. When such a function is passed to the `subPopSize` parameter of a mating scheme, it will be called to determine the population size of the offspring population at each generation.

A demographic function can accept one or both parameters `gen` and `pop`. Because simuPOP uses parameter names to determine the types of input parameters, other parameter names are unacceptable for this function. For example, an instant expansion model can be implemented using the following demographic function:

```
def ins_expansion(gen):
    'An instant population growth model'
    if gen < G0:
```

```
        return N0
    else:
        return N1
```

More complex demographic models such as an exponential population expansion model could be defined similarly. However, because a demographic function does not accept parameters other than `pop` and `gen`, variables such as `N0`, `G0`, and `N1` must be defined outside of function `ins_expansion`, either as global variables or in a namespace where this function is defined. This method is not recommended because functions defined in this way are not well encapsulated and cannot be reused in other scripts.

A better implementation is demonstrated in this example where a function `demoModel` is defined returning a demographic function with parameters `N0`, `G0`, and `N1`. This demographic function is used to determine the initial population size and is passed to the mating scheme to determine the size of the offspring generation at each generation. The change of population size is printed using two `Stat` and two `PyEval` operators that display parental and offspring population sizes at each generation.

SOURCE CODE 4.1 A Demographic Model with Population Split and Rapid Population Expansion

```
>>> import simuPOP as sim
>>> from math import ceil
>>> def demoModel(N0, N1, G0):
...     def func(gen):
...         if gen < G0:
...             return N0
...         else:
...             return N1
...     return func
...
>>> def simulate(demo, gen):
...     pop = sim.Population(size=demo(0))
...     pop.evolve(
...         initOps=sim.InitSex(),
...         preOps=[
...             sim.Stat(popSize=True),
...             sim.PyEval(r"'%d: %d ' % (gen, popSize)"),
...         ],
...         matingScheme=sim.RandomMating(subPopSize=demo),
...         postOps=[
...             sim.Stat(popSize=True),
...             sim.PyEval(r"'--> %d\n' % popSize"),
...         ],
...         gen=gen
...     )
...
>>> simulate(demoModel(100, 1000, 2), 5)
```

```
0: 100 --> 100
1: 100 --> 100
2: 100 --> 1000
3: 1000 --> 1000
4: 1000 --> 1000
>>>
```

Because population structure limits the free flow of genotype between subpopulations, we are also interested in the evolution of disease spectra in structured populations. In this chapter, we will use a model where the founder population split into m equally sized subpopulations right before population expansion. The split subpopulations will keep their relative proportion during population expansion. Example 4.2 defines a class `demoModel` to implement this demographic model. This class is defined in a module `reichDemo` and will be imported and used in later examples.

■ EXAMPLE 4.2

Although it is possible to split parental populations using operators such as `SplitSubPops` during evolution, it is easier to split populations in the demographic function. By passing parameter `pop` to the demographic function, the parental population will be passed to the demographic function, so member functions such as `Population.mergeSubPops` can be used to split, merge, and even resize the populations before mating starts. This example uses function `Population.splitSubPop` to split the population into m subpopulations at generation G_0 in both instant and exponential population expansion models.

If multiple demographic models are to be defined, an object-oriented implementation is recommended because it provides better readability and maintainability. This example defines a class `demoModel` that is initialized by parameters `N0`, `G0`, `N1`, `G1`, and `m`. These parameters are stored as attributes of the class to make them accessible to other member functions. An instant and an exponential population growth model are defined as member functions of this class. By defining a `self.__call__` member function, objects instantiated from the `demoModel` class are callable and can be used as a demographic function. For example, a `demoModel` object created from

```
demo_func = demoModel('exponential', 1000, 100000, 500, 500, 1)
```

can be used as a demographic function for an exponential population expansion model with $N_0 = 1000$, $N_1 = 100000$, $G_0 = 500$, and $G_1 = 500$ with no population structure.

SOURCE CODE 4.2 A Python Class That Defines Instant and Exponential Population Expansion Models

```
import math
class demoModel:
    def __init__(self, model, N0=1000, N1=100000, G0=500, G1=500, m=1):
        '''Return a demographic function with population split and expansion
        model: 'linear' or 'exponential'
        N0:    Initial population size.
        N1:    Ending population size.
        G0:    Length of burn-in stage.
        G1:    Length of population expansion stage.
        m:     Split population into m subpopulations before population expansion.
        '''
        self.N0, self.N1, self.G0, self.G1, self.m = N0, N1, G0, G1, m
        self.model = model

    def __call__(self, gen, pop=None):
        if self.model == 'instant':
            return self.ins_expansion(gen, pop)
        else:
            return self.exp_expansion(gen, pop)

    def ins_expansion(self, gen, pop):
        if gen < self.G0:
            return self.N0
        elif self.m > 1:
            if gen == self.G0:  # split population
                # avoid floating point problem
                pop.splitSubPop(0, [1./self.m] * self.m)
            return [self.N1//self.m]*(self.m-1)+[self.N1-self.N1//self.m*(self.m-1)]
        else:
            return self.N1

    def exp_expansion(self, gen, pop):
        rate = (math.log(self.N1) - math.log(self.N0))/self.G1
        if gen < self.G0:
            return self.N0
        elif self.m > 1 and gen == self.G0:  # split population
            pop.splitSubPop(0, [1. / self.m] * self.m)
        if gen == self.G0 + self.G1 - 1:
            N = self.N1
        else:
            N = int(self.N0 * math.exp((gen - self.G0) * rate))
        if self.m > 1:
            return [N // self.m] * (self.m-1) + [N - N // self.m * (self.m-1)]
        else:
            return N
```

4.2.2 Output Statistics

We are interested in two statistics: the total disease allele frequency and effective number of disease alleles at a locus, which measure the rareness and the allelic diversity of the disease alleles, respectively. Because allele frequencies add up to one, the total disease allele frequency can be obtained by $\sum_{i=1}^{\infty} f_i = 1 - f_0$. The observed effective number of disease alleles can be calculated from calculated allele frequencies using Equation 4.1. Example 4.3 demonstrates how to calculate these statistics in simuPOP.

■ EXAMPLE 4.3

simuPOP does not support the calculation of these statistics directly. However, the `Stat` operator provides native support for the calculation of allele frequency. We can calculate the total disease allele frequency and effective number of disease alleles at a locus from allele frequencies calculated by operator `Stat`. For example, the operator

```
Stat(alleleFreq=0),
PyEval('"%.2f" % (1-alleleFreq[0][0])'),
PyEval('"%.2f" % (1./sum([(alleleFreq[0][x]/(1-alleleFreq[0][0]))**2'
    'for x in alleleFreq[0].keys() if x != 0]))')
```

calculates and outputs both statistics at locus 0.

This approach, however, is not reusable because the expressions are valid for a specific locus and it would be a mess to use such expressions to output n_e for multiple loci or for n_e in multiple subpopulations. Moreover, this expression is error prone because it does not handle the case when there is no disease allele ($f_0 = 1$, $n_e = 0$). It makes sense to define a function to calculate n_e clearly.

This example defines a function `calcNe` to calculate the effective number of disease alleles at specified loci and save the result in a dictionary `ne` of the population's local namespace. This function can be passed to a `PyOperator` to calculate the effective number of disease alleles at specified loci. For example, operators

```
PyOperator(func=calcNe, param=[0, 1])
PyEval(r'"%.3f\t%.3f" % (ne[0], ne[1])')
```

would calculate and output this statistic at loci 0 and 1 during an evolutionary process.

This example calculates n_e at two loci. The first locus has 10 disease alleles with similar allele frequencies, and the second locus has a major locus that is 10 times more frequent than other 10 disease alleles. As expected from our previous analysis, the effective number of disease alleles at the second locus is smaller than that at the first locus.

SOURCE CODE 4.3 A Python Function to Calculate Effective Number of Alleles

```
>>> import simuPOP as sim
>>> def calcNe(pop, param):
...     'Calculated effective number of disease alleles at specified loci (param)'
...     sim.stat(pop, alleleFreq=param)
...     ne =
...     for loc in param:
...         freq = pop.dvars().alleleFreq[loc]
...         sumFreq = 1 - pop.dvars().alleleFreq[loc][0]
...         if sumFreq == 0:
...             ne[loc] = 0
...         else:
...             ne[loc] = 1. / sum([(freq[x]/sumFreq)**2 \
...                 for x in list(freq.keys()) if x != 0])
...     # save the result to the sim.Population.
...     pop.dvars().ne = ne
...     return True
...
>>> pop = sim.Population(1000, loci=2)
>>> sim.initGenotype(pop, freq=[0.9] + [0.01]*10, loci=0)
>>> sim.initGenotype(pop, freq=[0.9] + [0.05] + [0.005]*10, loci=1)
>>> calcNe(pop, param=[0,1])
True
>>> print(pop.dvars().ne)
0: 9.82506393861892, 1: 3.6730830927173557
```

When there are multiple subpopulations, we will need to calculate the effective number of disease alleles in each subpopulation as well as the whole population. It is challenging to calculate all these using a function, so Example 4.4 defines a customized operator. This operator is defined in module `reichStat` and will be used in later examples.

■ EXAMPLE 4.4

Although it is easy to extend function `calcNe` to perform more calculations according to parameters passed from a `PyOperator`, a more elegant solution is to define an operator with an interface that is similar to operator `Stat` so that it can be used as a regular simuPOP operator. The trick here is to define a class that is derived from operator `PyOperator` and point its `func` parameter to a member function of the class. Because this class is derived from `PyOperator`, it can accept regular operator parameters such as `begin`, `end`, `step`, and `at`, and call the member function when this operator is applied to a population.

This example defines a class `Ne` that is inherited from class `PyOperator`. The `__init__` function of this class accepts parameters `loci`, `subPops`, and `vars` that have the same default values and meanings as the corresponding parameters in a `Stat` operator. Function `self._Ne` will be called upon when this operator is applied to a population. This function calculates allele frequencies at specified loci and (virtual) subpopulations. If `'ne_sp'` is provided in parameter `vars`, allele frequencies in all or specified subpopulations will be calculated and saved in subpopulation dictionaries such as `subPop[(0,0)]['ne']`. A function `calcNe` is also defined in this module, which simply applies operator `Ne` to a population. Compared to the function defined in Example 4.3, this function accepts parameters `vars` and `subPops` and can calculate effective number of alleles in specified (virtual) subpopulations. Class `Ne` and function `calcNe` are defined in module `ch5_reichStat.py` and will be imported and used in later examples.

SOURCE CODE 4.4 A Self-Defined Operator to Calculate Effective Number of Alleles

```
import simuPOP as sim

class Ne(sim.PyOperator):
    '''An operator that calculates the effective number of disease alleles
    at specified loci.
    '''
    def __init__(self, loci=sim.ALL_AVAIL, subPops=sim.ALL_AVAIL, vars=[],
            *args, **kwargs):
        self.loci = loci
        self.vars = vars
        self.subPops = subPops
        sim.PyOperator.__init__(self, func=self._Ne, *args, **kwargs)

    def _calcNe(self, freq):
        'Calculate Ne from allele frequencies'
        if len(freq) == 0 or freq[0] == 1:
            return 0
        else:
            f_dis = 1 - freq[0]
            return 1. / sum([(freq[x]/f_dis)**2 \
                for x in list(freq.keys()) if x != 0])

    def _Ne(self, pop):
        # calculate allele frequency
        sim.stat(pop, alleleFreq=self.loci, subPops=self.subPops,
            vars=['alleleFreq_sp', 'alleleFreq'] if 'ne_sp' in self.vars else [])
        # determine loci
        loci = range(pop.totNumLoci()) if self.loci == sim.ALL_AVAIL else self.loci
        # ne for the whole population
        if len(self.vars) == 0 or 'ne' in self.vars:
            pop.dvars().ne = {}
            for loc in loci:
                pop.dvars().ne[loc] = self._calcNe(pop.dvars().alleleFreq[loc])
        if 'ne_sp' in self.vars:
            if self.subPops == sim.ALL_AVAIL:
                subPops = range(pop.numSubPop())
            else:
```

```
                subPops = self.subPops
            for sp in subPops:
                pop.dvars(sp).ne = 
                for loc in loci:
                    pop.dvars(sp).ne[loc]=self._calcNe(pop.dvars(sp).alleleFreq[loc])
        return True

def calcNe(pop, *args, **kwargs):
    'Calculate ne statistics of pop'
    Ne(*args, **kwargs).apply(pop)
```

4.2.3 Mutation Models

Reich and Lander [1] used an infinite allele model where there is a single wild-type allele and an infinite number of disease alleles. Because each mutation event introduces a new disease allele, no backward mutation is allowed. Because it is not feasible to simulate an infinite number of alleles, we use a k-allele mutation model to mimic an infinite allele mode. We assume that allele 0 is the wild-type allele and alleles 1 to $k-1$ are disease alleles. Because an allele in a k-allele mutation model can mutate to any other allele with equal probability $\frac{1}{k-1}$, the probability that a mutant duplicates with an existing allele will be low if k is large (e.g., $>10{,}000$). On the other hand, if some genes can only have a small number of variants, we can simulate them using k-allele models with smaller k values (e.g., $k=20$). Recurrent and backward mutations can no longer be ignored in these cases and it would be interesting to see if the theoretical model described in Section 4.1.2 still holds for these genes.

4.2.4 Multilocus Selection Models

To simplify computations, we follow Reich and Lander and assume that all disease alleles at a locus have the same fitness effect. That is to say, we categorize alleles at specified loci as wild-type (alleles A) and disease alleles (alleles a) and assign individual fitness values according to the number of disease alleles. For example, an additive single-locus selection model would assign fitness values 1, $1-s/2$, and $1-s$ for individuals with genotype AA, Aa, and aa, respectively.

For simplicity, we assume that loci under selection are unlinked, and there are no interaction among genotypes at these loci. In this case, the multilocus selection model can be treated essentially as a series of single-locus models. Although a general multilocus model of natural selection for diploid populations requires 3^L parameters to determine individual fitness from all combinations of genotypes at L loci, only $2L$ parameters (h_i and

s_i for each locus) are needed if we assume that natural selection works independently on each locus.

Although we can calculate fitness values at each locus, an individual in a population can only have one fitness value, so an overall fitness value has to be synthesized from fitness values at these loci. A big question, however, is how to combine locus-specific fitness values and whether the choice of multilocus selection models will have a significant impact on the outcome of the evolutionary process.

Several multilocus selection models are available. In addition to a multiplicative model where locus-specific fitness values are multiplied ($f = \Pi f_i$, where f_i is the fitness value at locus i and f is the overall fitness value of an individual), an additive model ($f = \max(0, 1 - \sum(1 - f_i))$) and an exponential model ($f = \exp(-\sum(1 - f_i))$) are also frequently used. Fortunately, despite the different methods to parameterize single-locus fitness values, the evolution of a locus in these multilocus selection models does not have to depend on interactions with other loci. Given the assumption that the marginal fitness value at a locus is g_{NN}, g_{NS}, and g_{SS} for genotypes NN, NS, and SS, respectively, and the frequencies and fitnesses of genotypes at other loci are p_i and f_i, respectively (where i iterates through all possible genotypes at other loci) for an individual having genotype XY at locus A (XY can be NN, NS, or SS), the expected (average) overall fitness value is

$$f'_{XY} = \sum_i p_i f_i f_{XY} = f_{XY} \sum_i p_i f_i \tag{4.5}$$

under the multiplicative multilocus fitness model and

$$f'_{XY} = \sum_i p_i (1 - (1 - f_i) - (1 - f_{XY})) = f_{XY} - 1 + \sum_i p_i f_i \tag{4.6}$$

under the additive model. From the viewpoint of locus A, because XY can be of any genotype at this locus, Equations 4.5 and 4.6 imply a systematic decrease ($\sum_i p_i f_i \leq \sum_i p_i = 1$, $1 - \sum_i p_i f_i \geq 0$) of fitness of *every* individual compared to a selection model with A as the only locus. Because f_{XY} are relative fitness values, such changes of fitness will have very little, if any, impact on the selection process. The probability of being selected is the same for a systematic multiplicative change of fitness values ($\frac{cg_{XY}}{\sum_k cg_{X_kY_k}} = \frac{g_{XY}}{\sum_k g_{X_kY_k}}$ where c is the multiplicative factor and k iterates over the individuals in the population considered) and is only slightly different for an additive change when $c = 1 - \sum_i p_i g_i \ll g_{XY}$ (usually the case for common diseases). Note that Equations 4.5 and 4.6 hold only when locus A and other loci are unlinked so that the frequency of individuals with certain genotypes can be written as the product of genotype frequencies at each locus.

■ EXAMPLE 4.5

If there is only one disease-predisposing locus, we can use a multiallelic selector (`MaSelector`) to specify individual fitness values according to the number of disease alleles they carry. This selector categorizes alleles at specified loci as wild-type (alleles A) and disease alleles (alleles a) and assign individual fitness values using a list of three elements (and 3^L elements if there are L disease predisposing loci). For example, operator

```
MaSelector(loci=0, wildtype=0, fitness=[1, 1-s/2., 1-s])
```

simulates a selection model that assigns fitness values 1, $1 - s/2$, and $1 - s$ for individuals with genotype AA, Aa, and aa at locus 0, respectively.

Multiple single-locus selection models can be combined using a `MlSelector` operator. For example, operator

```
MlSelector([
  MapSelector(loci=0, fitness=(0,0):1, (0,1):0.99, (1,1):0.98),
  MapSelector(loci=1, fitness=(0,0):1, (0,1):1, (1,1):0.98)
], mode=MULTIPLICATIVE)
```

simulates a two-locus selection model where individual fitness is the product of fitness values at two loci. That is to say, an individual with genotype `(0,1)` at locus `0` and `(1,1)` at locus `1` will have fitness value $0.99 \times 0.98 = 0.9702$.

This example demonstrates the independent evolution of loci under these multilocus selection models. It evolves three populations under different selection models: a single-locus additive model, a single-locus recessive model, and a multilocus selection model that is the production of these two models. As shown in this example, the allele frequencies of the two loci of the third population (last two columns of the output) roughly follow the loci in the first two population under corresponding single-locus selection models (first two columns).

SOURCE CODE 4.5 Simulation of Multiple Independent Single-Locus Selection Models

```
>>> import simuOpt
>>> simuOpt.setOptions(quiet=True, alleleType='binary')
>>> import simuPOP as sim
```

```
>>>
>>> pop = sim.Population(size=[10000]*3, loci=[1]*2, infoFields='fitness')
>>> simu = sim.Simulator(pop, rep=3)
>>> simu.evolve(
...     initOps=[
...         sim.InitSex(),
...         sim.InitGenotype(freq=[0.5, 0.5])
...     ],
...     preOps=[
...         sim.MaSelector(loci=0, fitness=[1, 0.99, 0.98], reps=0),
...         sim.MaSelector(loci=0, fitness=[1, 1, 0.99], reps=1),
...         sim.MlSelector([
...             sim.MaSelector(loci=0, fitness=[1, 0.99, 0.98]),
...             sim.MaSelector(loci=1, fitness=[1, 1, 0.99])],
...             mode=sim.MULTIPLICATIVE, reps=2)
...     ],
...     matingScheme=sim.RandomMating(),
...     postOps=[
...         sim.Stat(alleleFreq=[0,1], step=50),
...         sim.PyEval(r"'%.3f\t' % alleleFreq[0][1]", reps=0, step=50),
...         sim.PyEval(r"'%.3f\t' % alleleFreq[0][1]", reps=1, step=50),
...         sim.PyEval(r"'%.3f\t%.3f\n' % (alleleFreq[0][1], alleleFreq[1][1])",
...             reps=2, step=50),
...     ],
...     gen = 151
... )
0.496 0.499 0.499 0.496
0.358 0.435 0.367 0.439
0.242 0.385 0.264 0.382
0.172 0.345 0.186 0.328
(151, 151, 151)
>>>
```

4.2.5 Evolve!

With appropriate operators to perform mutation, selection, and output statistics, it is relatively easy to set up a simulation. Example 4.6 defines a function `evolvePop` to perform all the simulations. For the sake of simplicity, we do not provide a parameter handling mechanism in this example and will import this module and call upon function `evolvePop` directly for all simulations that will be discussed in the rest of this chapter.

■ EXAMPLE 4.6

This example defines a function `evolvePop` to evolve a population with one or more unlinked loci, using an instant or exponential population expansion demographic models, a k-allele mutation model with varying k and mutation rates, and a multiplicative multilocus model of natural selection.

This function uses the length of fitness values ($3L$) to determine the number of loci (L). It creates a demographic object and uses it to determine the size of an initial population (`demo_func(0)`). Every individual in this population has L unlinked loci and two information fields `fitness` and `migrate_to`, which are used for natural selection and migration, respectively. Before evolution, the population is initialized with random sex and random genotype with specified initial allelic spectra.

Mutation, natural selection (assignment of fitness values), and migration are applied to parental populations at the beginning of each generation before a random mating scheme is used to produce offspring populations. A `MlSelector` operator is used to combine fitness values at multiple loci using a multiplicative multilocus selection model. In order to accommodate arbitrary single-locus selection models, we accept three fitness values for each locus, so a list of $3L$ values need to be specified when there are L disease predisposing loci.

Because the calculation of population statistics can be time consuming, we wrap operators `Ne` and `PyEval` in operators `IfElse` so that statistics in the whole population and in each subpopulation are calculated only if valid output files are specified. Because expressions to output disease allele frequency and effective number of disease alleles are variable because of varying number of disease loci, they are constructed at the beginning of this function. The `IfElse` operators also accept additional keyword arguments so that parameters such as `begin` and `step` can be used to control when and at which frequency these statistics are calculated. This function calculates and returns to total disease allele frequency and effective number of disease alleles at all loci at the end of the evolution.

SOURCE CODE 4.6 Evolve a Population Subject to Mutation and Selection

```
import simuOpt
simuOpt.setOptions(quiet=True, alleleType='long')
import simuPOP as sim
from reichDemo import demoModel
from reichStat import Ne
from simuPOP.utils import migrIslandRates
from itertools import product

def evolvePop(model, N0, N1, G0, G1, initSpec, mu, k, fitness,
        m=1, migrRate=0, logfile='', sp_logfile='', **kwargs):
    '''Evolve a population with specified allele frequencies (parameter
    initSpec) using given demographic (model, N0, N1, G0, G1, m), mutation
    (a k-allele model with parameters mu and k) and natural selection models
    (a multi-locus selection model with fitness vector s). Total disease
    allele frequency and effective number of alleles in the population
    and in all subpopulations are recorded if names of log files are provided.
    This function returns a tuple of these two statistics at the end of the
    evolution. Additional keyword arguments could be used to control when and
    how often statisitcs are outputed.
    '''
```

```
L = len(fitness) // 3
if not hasattr(mu, '__iter__'): # if a single mutation rate is given
    mu = [mu]*L
# Create expressions to output f_e and ne at all loci, which are
#    "%d\t%.4f\t%.4f\n" % (gen, 1-alleleFreq[x][0], ne[x])
# for locus x.
statExpr = '"%d' + r'\t%.4f\t%.4f'*L + r'\n" % (gen,' + \
    ', '.join(['1-alleleFreq[%d][0], ne[%d]' % (x, x) for x in range(L)]) + ')'
demo_func = demoModel(model, N0, N1, G0, G1, m)
pop = sim.Population(size=demo_func(0), loci=[1]*L,
        infoFields=['fitness', 'migrate_to'])
pop.evolve(
    initOps=[
        sim.InitSex(),
        sim.InitGenotype(freq=initSpec)
    ],
    preOps=[
        sim.KAlleleMutator(k=k, rates=mu, loci=range(L)),
        sim.MlSelector([
            sim.MaSelector(loci=i, fitness=fitness[3*i:3*(i+1)])
            for i in range(L)], mode=sim.MULTIPLICATIVE),
        sim.Migrator(rate=migrIslandRates(migrRate, m), begin=G0+1),
    ],
    matingScheme=sim.RandomMating(subPopSize=demo_func),
    postOps=[
        sim.IfElse(logfile != '' or sp_logfile != '',
            Ne(loci=sim.ALL_AVAIL, vars=['ne'] if m == 1 else ['ne', 'ne_sp']),
            **kwargs),
        sim.IfElse(logfile != '',
            sim.PyEval(statExpr, output='>>' + logfile), **kwargs),
        sim.IfElse(m > 1 and sp_logfile != '',
            sim.PyEval(statExpr, output='>>' + sp_logfile,
            # subPops=sim.ALL_AVAIL will evalulate the expression in each
            # subpopulation's local namespace (vars(sp)).
            subPops=sim.ALL_AVAIL, begin=G0), **kwargs),
        ],
    finalOps=Ne(loci=sim.ALL_AVAIL),
    gen = G0 + G1
)
return tuple([1-pop.dvars().alleleFreq[x][0] for x in range(L)] + \
    [pop.dvars().ne[x] for x in range(L)])
```

4.2.6 Validation of Theoretical Results

To verify theoretical estimates of equilibrium effective number of disease alleles, we used different combinations of N, μ, and s and evolved constant size populations for a long period of time until they reaches mutation, selection and drift equilibrium. Total disease allele frequency and effective number of disease alleles at the end of the simulation are collected and plotted in Figure 4.2.a and b, along with their theoretical expectations using Equations 4.3 and 4.2, respectively. Note that curves in Figure 4.2.b are almost horizontal, indicating that the effective numbers of alleles for rare and common diseases are similar in equilibrium states. This is because the

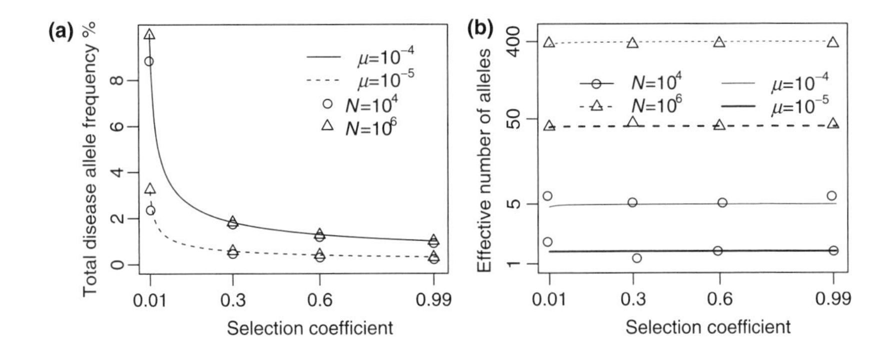

FIGURE 4.2 Allele frequency and effective number of alleles of simple recessive diseases in equilibrium state. Total disease allele frequency (a) and effective number of disease alleles (b) of simple recessive diseases in equilibrium state, using different parameter settings ($N = 10^4$ or 10^6, $\mu = 10^{-4}$ or 10^{-5}, $s = 0.01, 0.3, 0.6, 0.99$). Solid and dotted lines are theoretical expectations of $f_e = \sqrt{\frac{\mu}{s}}$ for $\mu = 10^{-4}$ and 10^{-5}, respectively. Each dot in the figures represents average values of 10 replicate simulations.

equilibrium n_e is determined by N, μ, and $f_0 = 1 - \sum_{i>1} f_i$ (Equation 4.2) and the only differentiating factor f_0 falls largely between 0.9 and 1 for these diseases and have only a small impact on the value of equilibrium n_e.

The evolution of n_e under an instant population expansion model also conforms well with theoretical estimates. For example, we picked three diseases with different initial total disease allele frequencies and evolved them using a demographic model where the founder population grows instantly from $N_0 = 10^4$ to $N_1 = 10^7$. The dynamics of the effective number of disease alleles of three diseases are plotted in Figure 4.3, with theoretical estimates calculated according to Equation 4.4 plotted in dotted lines. It is clear that the effective number of disease alleles increases slower for diseases with higher equilibrium total disease allele frequencies.

4.3 EXTENSIONS TO THE BASIC MODEL

4.3.1 Impact of Demographic Models

Different human populations have different demographic histories. Some populations like the isolated Scandinavian Saami have had approximately constant population size, but most human populations are thought to have undergone a rapid population expansion [10]. Because isolated populations

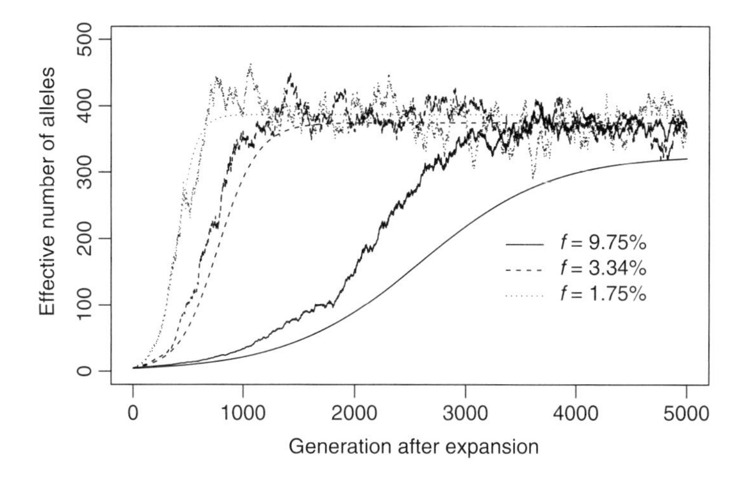

FIGURE 4.3 Evolution of simple recessive diseases. The dynamics of the effective number at instant population expansion from $N_0 = 10^4$ to $N_1 = 10^7$ for two diseases with equilibrium total disease allele frequencies, which are determined from $f_e = \sqrt{\frac{\mu}{s}}$ where $\mu = 10^{-5}$ and s =0.01, 0.99, respectively.

generally have small population sizes, genetic diseases in these populations have a relatively small number of alleles (see Equation 4.2). Consequently, we are only interested in large populations with high equilibrium effective number of disease alleles.

Among many population expansion models, the exponential population growth model is a simple but yet reasonably realistic one. It is widely assumed that the general human population had constant size $N_0 = 10,000$ until $G_0 = 5000$ generations before the present, and then expanded exponentially to its present-day size of $N_1 = 6$ billion [11–13]. $G_0 = 5000$ corresponds to 100,000 years ago given a 20 year generation time. Such a huge population is difficult to simulate using simulations. Fortunately, because the human populations have complex population structures and are far from randomly mating, their effective population sizes are much smaller than their census population sizes. We use $N_1 = 10^6$ or $N_1 = 10^7$ as the size of the present population in our simulations.

The choice of demographic models can have a large impact on the evolution of allelic spectra. As pointed out in Ref. [1], slower population expansion would result in a slower growth in allelic diversity. If we use an exponential population growth model, the human population increases rather slowly most of the time. This has two consequences: during the slow-growing period, small population size tends to limit the growth of the effective number of alleles, so n_e will increase more slowly than in a faster

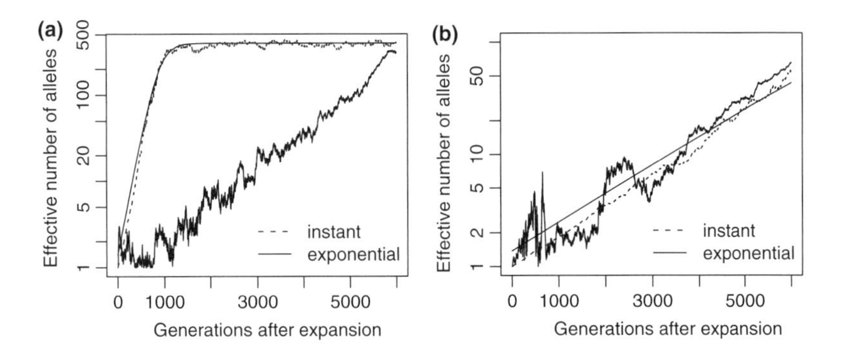

FIGURE 4.4 Impact of demographic models on effective number of alleles. Evolution of n_e of a rare ($s = 0.99$, (a)) and a common disease ($s = 0.01$, (b)) under instant and exponential population growth models. $N_0 = 10^4$, $N_1 = 10^7$, and $\mu = 10^{-5}$. The two solid black curves are theoretical estimates under the infinite allele model and instant growth.

growth model; the "large population" stage is effectively shorter than in the instant growth model and gives diseases less time to reach equilibrium.

Figure 4.4 plots the dynamics of n_e in four simulations, which use the same basic parameters ($N_0 = 10^4$, $N_1 = 10^7$, $\mu = 10^{-5}$) but different selection coefficients ($s = 0.99$ for rare diseases and $s = 0.01$ for common diseases) and demographic models (instantaneous, exponential). Although equilibrium n_e is close to 400 for all six cases, the kinetics are quite different. The effective number of alleles for common diseases increases more slowly than that of the rare disease, but the difference is smaller for the exponential growth model than that of the instant growth model. Due to the demographic difference, n_e of a rare disease under the exponential growth model at generation 5000 is at the same level as the instant population growth model at generation 1000.

4.3.2 Impact of the Mutation Model

The theoretical model uses an infinite allele mutation model. When the effective population size is large, this model leads to an unrealistically large n_e. For example, when $N = 10^9$ and $\mu = 10^{-5}$, the equilibrium $n_e = 3.2 \times 10^4$ for a common disease with a total disease allele frequency of 0.2. However, due to the constraints on gene length, silent or recurrent mutations, effective number of alleles for real human diseases is usually smaller than this number.

In the previous simulations, we used a k-allele model with a large number of alleles ($k > 10^4$) to mimic the infinite allele model. Recurrent mutations

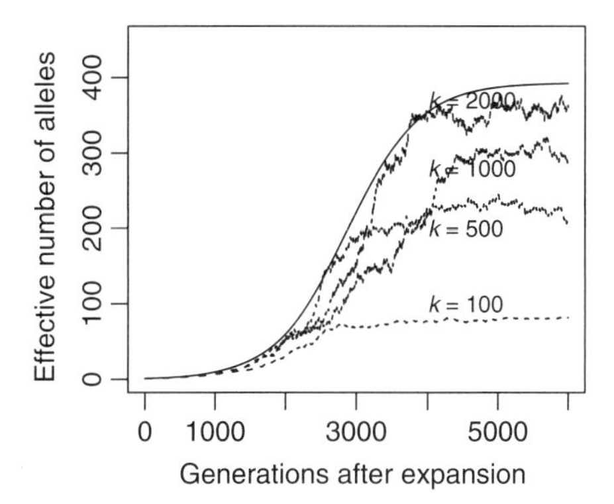

FIGURE 4.5 Impact of mutation models on effective number of alleles. Change of n_e with number of allelic states 100, 500, 1000, 2000, using an instant population growth model. $N_0 = 10^4$, $N_1 = 10^7$, $\mu = 10^{-5}$, $s = 0.1$. Solid curve is the theoretical estimate under the infinite allele model.

do occur, but at such a small rate that they have almost no impact on the proportion of alleles derived from before population expansion or on the equilibrium effective number of alleles.

The probability of recurrent mutation increases with decreasing k. This leads to smaller observed equilibrium n_e compared to n_e under the infinite allele model. To verify this, we simulate the evolution of n_e of a disease with $s = 0.1$ in a population that has grown instantly from $N_0 = 10^4$ to $N_1 = 10^7$ at 6000 generations ago, using k-allele models with $k = 100, 500, 1000,$ or 2000. Among these k-values, only $k = 2000$ reaches the n_e expected under the infinite allele model (Figure 4.5). Although it is not clear how exactly n_e will change with k when $k > n_e$, when $k < n_e$ and there are enough disease alleles to fill every allelic state (high N, μ, or small s), n_e will be close to k ($n_e \sim \left((k-1)\left(\frac{1}{k-1}\right)^2\right)^{-1} = k - 1$) regardless of the exact values of N, μ, or s ($k = 100$ in Figure 4.5).

Although equilibrium n_e with small maximum number of allele states is smaller than that of the infinite allele model, at the beginning the former increases faster than the latter. This takes place because μ_N, the mutation rate from susceptibility to normal alleles, is no longer negligible and accelerates the dissolution of the dominant disease allele when k is small.

4.3.3 Impact of Subpopulation Structure

The human population went through complex migration patterns that might impact the allelic spectra of human diseases. We will start from the simplest cases when no migration is allowed among subpopulations.

Suppose the population after instant expansion is split into m equally sized subpopulations, which then evolve independently without migration afterward. In each subpopulation, the equilibrium n_e equals $1 + 4\frac{N}{m}\mu(1 - f_e)$ (Equation 4.1 with population size replaced by $\frac{N}{m}$) where f_0 is assumed to be the same in all subpopulations because it is determined by the nature of disease. Because of the smaller population size and expected effective number of alleles, subpopulations reach equilibrium state quicker than in a large uniform population. Consequently, n_e of a structured population tends to evolve faster than that in a uniform population and equilibrium n_e is also larger.

The equilibrium n_e in the whole population is located between $n_l = 1 + 4\frac{N}{m}\mu(1 - f_e) \sim n_e/m$ (when the allelic spectra are identical in all subpopulations) and $n_h = \left(\sum_{i=1}^{m}\sum_{j=1}^{n_j}\left(\frac{mf_j}{mf_e}\right)^2\right)^{-1} = \left(\sum_{i=1}^{m}\frac{1}{m^2}n_l\right)^{-1} = mn_l = m + 4N\mu(1 - f_e) \sim m + n_e$ (when disease alleles are totally different among subpopulations, here we assume equal f_e and n_l in all subpopulations). Assuming a split-and-grow demographic model, allelic spectra in subpopulations are similar at the beginning and become increasingly distinct over time. Therefore, n_e will approach n_h in the long run when the differences between allelic spectra in subpopulations increase with time. The difference between n_h and n_e in a single population is determined by the number of subpopulations m. For example, in the case of a rare disease in many small tribes, each tribe may be dominated by one or a few tribe-specific mutants. The overall n_e will be close to the number of tribes, larger than the small n_e in individual tribes.

To confirm these analyses, we evolve a rare ($s = 0.9$) and a common ($s = 0.01$) disease, using a demography where a founder population is expanded instantly from $N_0 = 10^4$ to $N_1 = 10^7$ and at the same time splits into m ($m = 20$) subpopulations. The equilibrium effective number of alleles of the whole population is around 400 for both diseases if we ignore population structure (Figure 4.6.a and b). We see that n_e in each subpopulation evolves roughly as expected, and the overall n_e, as the result of composition of allelic spectra in 100 subpopulations, increases faster and arrives at a larger equilibrium n_e than that expected theoretically using a single population.

4.3.4 Impact of Migration

Although n_e in a structured population tends to be larger than in a single population, n_e in each subpopulation evolves as expected, unless new

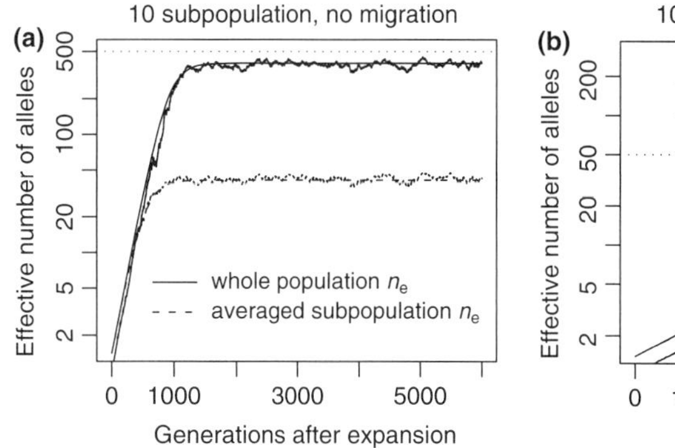

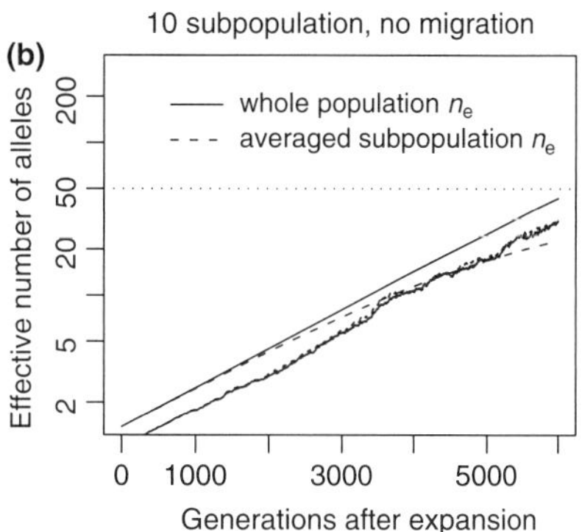

FIGURE 4.6 Impact of population substructure on effective number of alleles. Evolution of n_e of a rare ($s = 0.99$, (a)) and a common disease ($s = 0.01$, (b)) after instant population expansion from $N_0 = 10^4$ to $N_1 = 10^7$ with $\mu = 10^{-5}$. The population is split into 10 subpopulations after expansion. No migration is allowed. Thick lines are total n_e, thin lines are n_e of the first subpopulation. Note that the y-axes are in log scale.

mutants are introduced by migration. From a single subpopulation point of view, migration is a way to introduce new mutants, usually at a higher intensity than by mutation alone. Consequently, in a subpopulation with migration, n_e is larger than that in an isolated subpopulation. On the other hand, while the homogenizing effect of migration is not obvious soon after population split, when allelic spectra in subpopulations are similar to each other, it mixes alleles from subpopulations and keep n_e of a structured population away from n_h. In an extreme scenario when migration is so strong that all subpopulations have the same allelic spectra, n_e of the whole population is the same as that of a single subpopulation.

Migration does not have the same impact on common and rare diseases. When a disease is common, a significant proportion of migrants are affected. The impact of migration on the allelic spectrum is strong compared to weak mutation and selection. When a disease is rare, there are few affected migrants, so disease alleles tend to remain private in their own subpopulation. Since selection is strong in this case, migration is no longer a dominating force.

These analyses are confirmed by Figure 4.7, which is similar to Figure 4.6 except that migration is allowed between subpopulations. In these simulations, 0.1% of individuals in a subpopulation migrate to the adjacent subpopulations at each generation. When a disease is common, migration is strong enough to make allelic spectra more similar in all subpopulations.

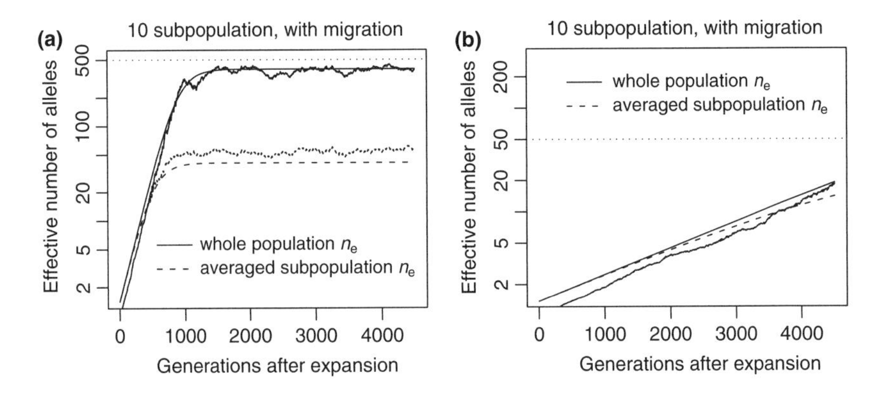

FIGURE 4.7 Impact of migration on the evolution of effective number of alleles. Evolution of n_e of a rare ($s = 0.99$, (a)) and a common disease ($s = 0.01$, (b)) after instant population expansion from $N_0 = 10^4$ to $N_1 = 10^7$ with $\mu = 10^{-5}$. The population is split into 10 subpopulations after expansion. Migration following an island model with a migration rate of 0.1% is allowed. Thick lines are total n_e, thin lines are n_e of the first subpopulation. Note that the y-axes are in log scale.

The allelic spectrum of the whole population is therefore closer to those of the subpopulations (compare common diseases in Figure 4.6b).

In conclusion, the allelic structure is more diverse in a subpopulation with new mutants introduced as a result of migration than the allelic structure of an isolated subpopulation. However, from the whole population point of view, the homogenizing effect of migration decreases n_e, which otherwise is greater than n_e in a single population. The impact depends on the number of subpopulations, level of migration, and the commonness of the disease.

4.3.5 Distribution of Equilibrium Disease Allele Frequency

Although the disease allele frequencies (f_e) under a mutation selection equilibrium can be estimated for a population of infinite size, the actual distribution of disease allele frequencies in finite populations varies. Because f_e of loci under strong purifying selection is small (e.g., $f_e = \sqrt{\frac{\mu}{s}}$ for the case of a recessive disease when $s \gg \mu$), f_e can rarely reach higher allele frequency. On the other hand, for common alleles with $s \sim \mu$, the impact of random genetic drift is stronger, which leads to a wider distribution of f_e. Assuming forward and reverse mutation rates μ_S, μ_N, and selection coefficient s for an additive model with fitness values 1, $1 - s/2$, and $1 - s$ for genotypes `00`, `01`, and `11`, respectively, the distribution of equilibrium overall frequency f_e of susceptibility alleles in the population is given by Wright's formula:

$$f(f_e) = cf_e^{(\beta_s - 1)} (1 - f_e)^{(\beta_N - 1)} e^{\sigma(1-f_e)} \tag{4.7}$$

where $\beta_S = 4N\mu_S$, $\beta_N = 4N\mu_N$, and $\sigma = 2Ns$ (s in this book is twice that is Ref.) are scaled parameters. The normalization constant c can be obtained by numerical integration. This formula works best in the cases of weak selection $\left(\text{e.g., } s \le 10^{-3}\right)$. For larger s $\left(\text{e.g., } s = 0.2,\ N = 10^4, \sigma = 4000\right)$, the exponential term will dominate $f(f_e)$ and make it essentially a delta function at 0.

To observe the distribution of f_e in finite populations, we evolved 50 populations for extended generations, using a fixed population size of $N = 10^5$, $\mu = 10^{-5}$, and different selection coefficients. We use a k-allele model with $k = 200$, so the forward and backward mutation coefficients are $\beta_S = 4\mu N = 4$ and $\beta_N = 0.02$, respectively. The total disease allele frequencies and effective numbers of alleles of these simulations are plotted in Figure 4.8.

4.3.6 Varying Selection and Mutation Coefficients

In this section, we study the allelic spectra of DSL responsible for polygenic diseases. Because we would like to know if our results for single-locus selection models still holds for multilocus cases, we start from multilocus selection models where natural selection is assumed to act on these loci independently. More specifically, we assume that the disease has L DSL located on different chromosomes and the overall fitness fits a multiplicative or additive multilocus model.

DSL of polygenic diseases usually do not contribute equally to the diseases. There might be some DSL that are under strong selection and many other DSL that are only slightly deleterious. The same holds for mutation rates. These locus-by-locus differences may cause interactions among DSL and disallow the dissection of the multilocus model into several single-locus models.

Figure 4.9 shows the results of two simulations with varying selection coefficients (a) or varying mutation rates (b). The diseases are recessive at each DSL and the overall fitness is modeled by a multiplicative model. The population size and mutation rate are $N = 10^5$ and $\mu = 10^{-5}$, respectively. Simple estimates of f_0 and n_e assuming a single-locus model are given by the solid lines, which match f_0 and n_e at each DSL of the multilocus model almost perfectly. This is true for other simulations we have run using additive single and multilocus models (results not shown). It is therefore

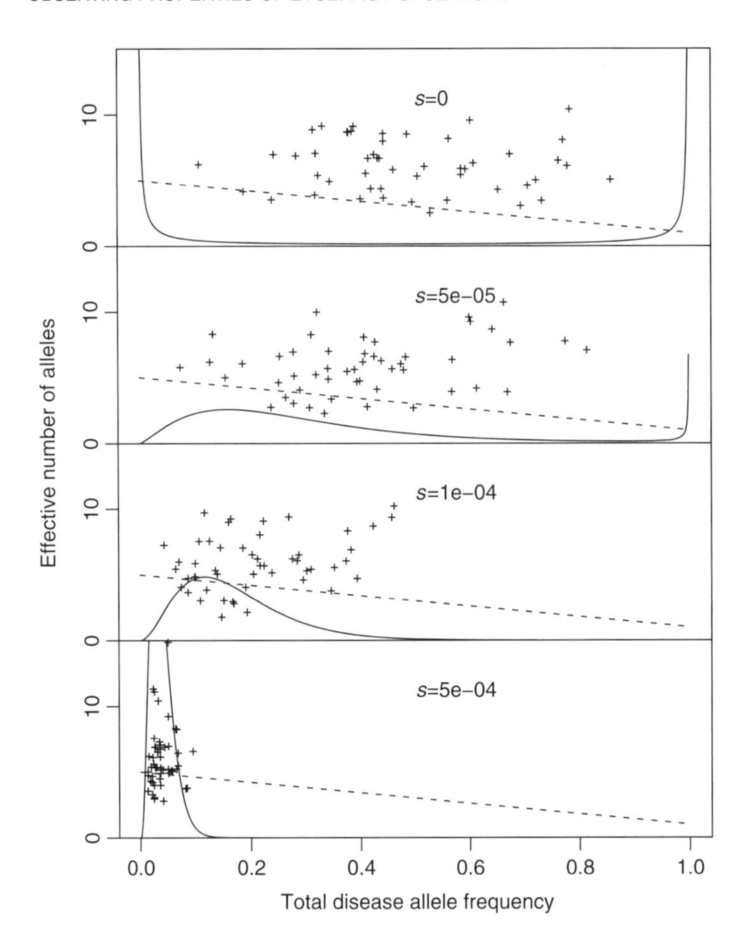

FIGURE 4.8 Distribution of total disease allele frequencies. Total disease allele frequency and effective number of alleles of 50 simulations with $N = 10^5$, $\mu = 10^{-5}$, $k = 100$, $s = 0.5 \times 10^{-4}, 10^{-4}, 0.5 \times 10^{-3}$ (from top to bottom). Solid curves are densities given by Equation 4.7. The mean $1 - f_0$ of all DSL at the last generation is used to estimate μ_S and μ_N.

safe to treat this multilocus model as a set of independent single-locus models.

4.3.7 Evolution of Disease Predisposing Loci Under Weak Selection

Evolution of the allelic spectrum of a DSL is determined by population size, mutation rate, and most importantly by the total disease allele frequency of the DSL. In the context of a common disease, the distribution of f_0 is

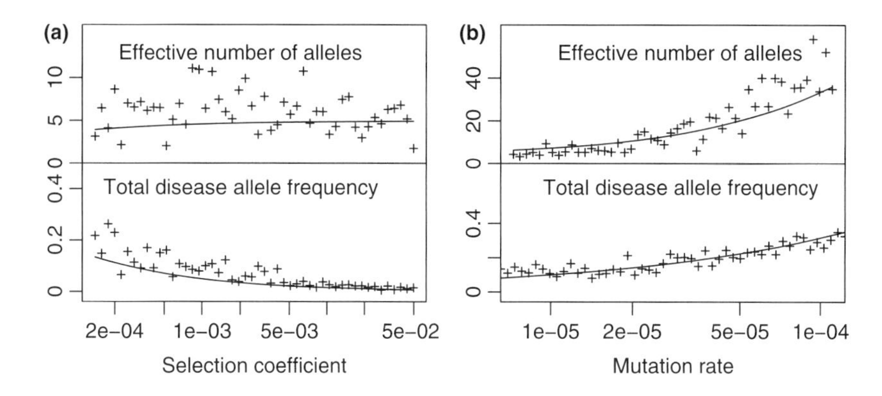

FIGURE 4.9 Varying selection and mutation coefficients. Total disease allele frequency (f_0) and effective number of alleles (n_e) of a common disease caused by 50 loci with varying selection coefficient (a) or mutation rate (b) in a population of constant size 10^5. (a) Disease with varying selection coefficient at each DSL. DSL are equally spaced in the interval $(\ln 0.0001, \ln 0.05)$ with $\mu = 10^{-5}$ (b) Disease with varying mutation rate at each DSL that are equally spaced in the interval $(\ln 6.5 \times 10^{-6}, \ln 1.2 \times 10^{-4})$ with $\mu = 0.01$. Diseases are recessive at each DSL. Solid lines are theoretical estimates given by $f_0 = \sqrt{\frac{\mu}{s}}$ and Equation 4.2.

quite dispersed, especially when N is small (see Figure 4.8). Consequently, we would expect highly dispersed f_{exp} at the beginning of population expansion. This results in different evolutionary patterns between DSL with identical parameter settings. An example in the supplementary material confirms this. In this example, four DSL of a polygenic disease evolve under identical parameter settings. The evolution of n_e at these DSL differs greatly because their f_{exp} deviate from f_0 due to small initial population size and the resulting dispersed distribution of f_0.

Figure 4.10 plots the dynamics of the total disease allele frequency (f) and the effective number of alleles (n_e) of four DSL of a polygenic disease ($L = 10$), under an instant population growth model with $N_0 = 10^4$, $N_1 = 10^6$, and $G_0 = 5000$, with $\mu = 10^{-5}$ and $s = 0.001$ identical at all DSL. Multiplicative multilocus selection model is used. Although the equilibrium f_0 of all DSL is 0.1, the distribution of f_0 at generation 5000 is quite dispersed because of the small population size during the burn-in period. After population expansion, f_0 of all DSL approaches 0.1 slowly, but the evolution of n_e at each DSL is roughly determined by the total disease allele frequency of DSL at the beginning of the population expansion (see the theoretical curves in the right panel of Figure 4.10). Consequently, n_e of the DSL of a common disease may approach their equilibrium states at vastly different rates.

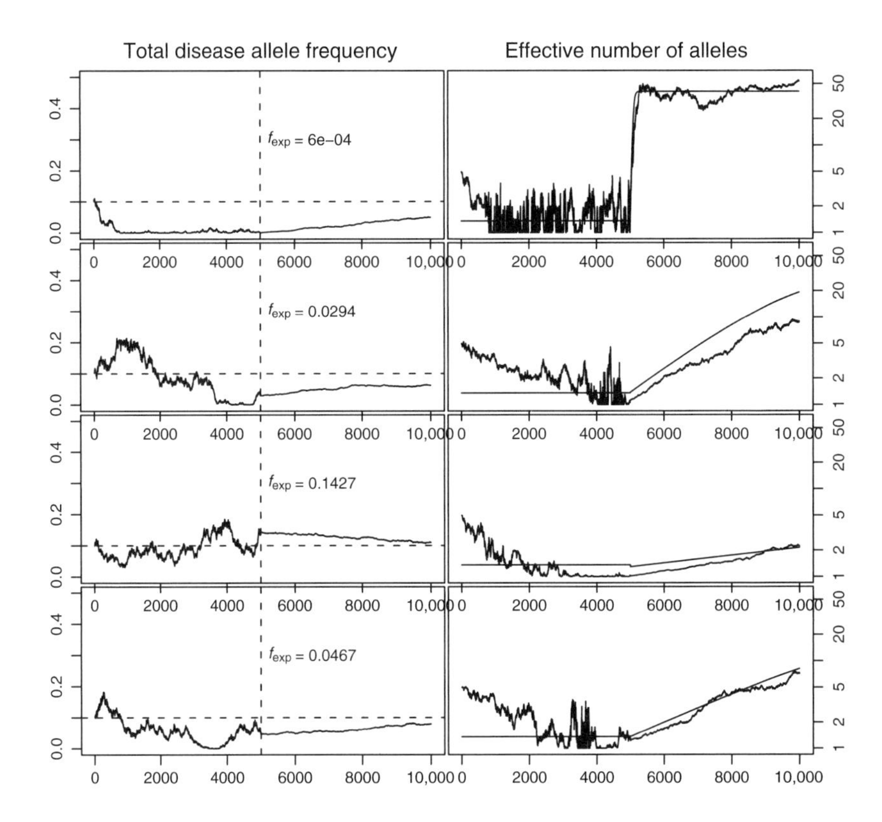

FIGURE 4.10 Evolution of disease susceptibility loci. Dynamics of the total disease allele frequency (left panel) and the effective number of alleles (right panel) of four DSL of a polygenic disease ($L = 10$), using a multiplicative multilocus selection model and an instant growth demographic model with $N_0 = 10^4$, $N_1 = 10^6$, $G_0 = G_1 = 5000$, $s = 0.001$, and $\mu = 10^{-5}$ are identical for all DSL. The horizontal dashed lines in the left panel are $f_0 = 0.1$, which is the equilibrium total disease allele frequency estimated using a single-locus model. Solid lines in the right panel are theoretical estimate of n_e with f_0 being the total disease allele frequency of DSL just before population expansion.

4.3.8 Discussion

Summarizing our simulations, we expect increased allelic diversity with the following:

1. Larger population size (or larger recent population size for varying demographic model),
2. Higher mutation rate, more allele states (may be the result of a longer gene),

3. Smaller total disease allele frequency at the beginning of population expansion. This may be the result of higher selection pressure, shorter evolution time before expansion, or by chance (genetic drift),
4. Longer evolutionary history (older mutants), and
5. More subpopulations and/or lower migration among subpopulations.

We have run many simulations to study the impact of each genetic feature on the allelic diversity. For example, we used $k = 200, 500, 1000, 2000$ to see the impact of possible number of allelic states on the effective number of alleles. We also varied selection rate ($s \in [0.0001, 0.99]$) and mutation rate ($\mu \in \left[6.5 \times 10^{-6}, 1.2 \times 10^{-4}\right]$) over a wide range of possible values. Although direct comparison is not possible since these simulations use different population size, length of evolution, and so forth, we can conclude from all simulations that the allelic spectrum is most sensitive to the total disease allele frequency at the beginning of population expansion (f_{exp}). For other genetic features, it is difficult to rank their relative importance. If two conflicting forces (such as higher mutation rate but quicker population expansion) are involved, it seems necessary to resort to a simulation program.

The total disease allele frequency at the disease locus (or loci in a polygenic disease setting) at the beginning of population expansion has a great impact on the evolution of allelic diversity. Although disease alleles that are under strong purifying selection generally have small total disease allele frequencies, disease alleles that are under mild selection can reach higher frequencies, especially in small populations where genetic drift is strong. Because population size is a limiting factor of effective number of disease alleles, a severe bottleneck would remove most alleles of both rare and common diseases and result in simple spectra. If a bottleneck is recent, both rare and common diseases will have simple spectra, while rare diseases will recover their diversity quicker than common ones.

The above independence-based argument, however, is potentially incomplete, because the relationship between commonness and weak selection for a polygenic disease does not have to hold as in the case of monogenic diseases. If a common disease is caused by several loci under weak selection, then according to our model, these DSL will be common and will have simple allelic spectra. This is the CDCV hypothesis in the cases of polygenic diseases [15]. However, a common disease may well be caused by rare alleles at numerous DSL, if each can single-handedly cause the disease. This is referred to as the genetic heterogeneity model. Our model suggests that these DSL will be rare and have highly diverse allelic spectra. Although theoretical studies and empirical data suggest that DSL for a complex disease are usually under weak selection, we cannot rule out the possibility of the heterogeneity model [16]. As a matter of fact, a common disease may be caused by a few loci with common alleles and many more with rare alleles. These common loci are most ready to be mapped.

REFERENCES

1. D. E. Reich and E. S. Lander, On the allelic spectrum of human disease. *Trends Genet*, 17(9):502–510, 2001.
2. J. K. Pritchard, Are rare variants responsible for susceptibility to complex diseases? *Am J Hum Genet*, 69(1):124–137, 2001.
3. B. Peng and M. Kimmel, Simulations provide support for the common disease-common variant hypothesis. *Genetics*, 175(2):763–776, 2007.
4. P. Tonin, B. Weber, K. Offit, F. Couch, T. R. Rebbeck, S. Neuhausen, A. K. Godwin, M. Daly, J. Wagner-Costalos, D. Berman, G. Grana, E. Fox, M. F. Kane, R. D. Kolodner, M. Krainer, D. A. Haber, J. P. Struewing, E. Warner, B. Rosen, C. Lerman, B. Peshkin, L. Norton, O. Serova, W. D. Foulkes, and J. E. Garber, Frequency of recurrent BRCA1 and BRCA2 mutations in Ashkenazi Jewish breast cancer families. *Nat Med*, 2(11):1179–1183, 1996.
5. E. S. Lander, The new genomics: global views of biology. *Science*, 274(5287):536–539, 1996.
6. N. S. Fearnhead, B. Winney, and W. F. Bodmer, Rare variant hypothesis for multifactorial inheritance: susceptibility to colorectal adenomas as a model. *Cell Cycle*, 4(4):521–525, 2005.
7. I. Gorlov, O. Gorlova, M. Frazier, M. Spitz, and C. Amos, Evolutionary evidence of the effect of rare variants on disease etiology. *Clin Genet*, 79(3):199–206, 2011.
8. M. Kimura and J. F. Crow, The number of alleles that can be maintained in a finite population. *Genetics*, 49:725–738, 1964.
9. D. L. Hartl and R. B. Campbell, Allele multiplicity in simple Mendelian disorders. *Am J Hum Genet*, 34(6):866–873, 1982.
10. M. Laan and S. Pääbo, Demographic history and linkage disequilibrium in human populations. *Nat Genet*, 17(4):435–438, 1997.
11. H. C. Harpending, M. A. Batzer, M. Gurven, L. B. Jorde, A. R. Rogers, and S. T. Sherry, Genetic traces of ancient demography. *Proc Natl Acad Sci USA*, 95(4):1961–1967, 1998.
12. L. Kruglyak, Prospects for whole-genome linkage disequilibrium mapping of common disease genes. *Nat Genet*, 22(2):139–144, 1999.

13. D. E. Reich and D. B. Goldstein, Detecting association in a case-control study while correcting for population stratification. *Genet Epidemiol*, 20(1):4–16, 2001.
14. J. K. Pritchard and M. Przeworski, Linkage disequilibrium in humans: models and data. *Am J Hum Genet*, 69(1):1–14, 2001.
15. D. J. Smith and A. J. Lusis, The allelic structure of common disease. *Hum Mol Genet*, 11(20):2455–2461, 2002.
16. Q. Yang, M. J. Khoury, J. Friedman, J. Little, and W. D. Flanders, How many genes underlie the occurrence of common complex diseases in the population? *Int J Epidemiol*, 34(5):1129–1137, 2005.

CHAPTER 5

SIMULATING POPULATIONS WITH COMPLEX HUMAN DISEASES

Complex diseases such as hypertension and diabetes are usually caused by multiple disease susceptibility genes, environmental factors, and interactions among them. Simulating populations or samples with complex diseases is an effective approach to study the likely genetic architecture of these diseases and to develop more efficient gene mapping methods. Compared to traditional backward-time (coalescent) methods, population-based, forward-time simulations are more suitable for this task because they can simulate almost arbitrary demographic and genetic features, especially complex models of natural selection. Forward-time simulations also allow the researchers to perform head-to-head comparisons among gene mapping methods based on different study designs and ascertainment methods. However, due to a number of methodological and computational constraints, forward-time simulations have not been widely used for the simulations of genetic samples. This chapter demonstrates a few methods to simulate populations with complex human diseases and discusses advantages and limitations of these simulation approaches.

Forward-time Population Genetics Simulations: Methods, Implementation, and Applications,
Bo Peng, Marek Kimmel, and Christopher I. Amos.

5.1 INTRODUCTION

Simulated data sets of known disease predisposing loci (DPL) have been widely used in the development and application of statistical methods that detect susceptibility genes for human genetic diseases [1, 2], for example, to compare the power of popular study designs and statistical methods for GWA studies [3–5], to simulate case–control samples for evaluating the power of new statistical methods [6], and to study the performance of statistical tests under different disease models [7, 8].

Many computer programs have been developed to simulate genetic data. If we exclude specialized methods that simulate genetic data for particular types of samples, currently available simulation methods can be categorized roughly into backward-time-based (coalescent), forward-time-based, and sideways methods [9].

Sideways methods permute or sample from existing genome sequences. Although they excel at retaining allele frequency and linkage disequilibrium (LD) information from existing sequences, they are limited in their ability to introduce new genetic features (such as the effects of natural selection) and new haplotypes.

Coalescent methods excel at simulating random samples, but it is difficult to use these methods to simulate casecontrol or other types of samples with genetic diseases because a coalescent simulation constructs a genealogical tree from samples with unknown genotypes and cannot effectively control the number of affected individuals once the genotypes of these samples are simulated. If a large number of samples are simulated before a disease model is applied, the coalescent method becomes inefficient, especially when long genomic sequences are simulated, unless special algorithms are used to approximate the standard coalescent process. In addition, many coalescent method-based programs simulate samples with random marker locations, which makes defining a genetic disease with consistent DPL difficult. Finally, most of these methods were designed to simulate case–control samples of relatively simple disease models and, therefore, have limited applicability to important research areas such as the detection of gene–environment interaction, admixture mapping, or family-based associations.

Forward-time simulation methods evolve a population forward in time, subject to arbitrary genetic and demographic factors. Because such a simulation can closely mimic the complex evolutionary histories of human populations that harbor the genetic diseases of interest, these methods can, in theory, simulate genetic samples with arbitrary complexity. Arbitrary disease models could be applied to the resulting population from which samples based on different study designs can be drawn and analyzed. However, this method suffers from a number of theoretical and computational issues, which have prevented the power of this simulation method from being fully explored in existing forward-time simulation methods:

- *Efficiency* The forward-time simulation method is inefficient because ancestors who do not have offspring in the resulting population are simulated, and a large population must be simulated before samples can be drawn from it. This is becoming less and less of a problem due to the increasing power of personal computers and the wider availability of high-performance cluster computer systems.
- *Simulation Length and Initialization* Unlike the coalescent approach, which starts from a single individual (MRCA), forward-time simulations usually start from an initial population of moderate size. How to initialize this population is a surprisingly difficult question. In addition, the properties of populations simulated using a forward-time approach depend heavily on the initial populations, which are often simulated under arbitrary equilibrium assumptions. Simulating an initial population inaccurately can affect the validity of conclusions made about the performance of a test.
- *Introduction of Disease* In the coalescent approach, the age of the mutant is random because the age of MRCA of any affected individual is random. This is difficult to achieve using a forward-time approach. A more serious problem is that newly introduced disease mutants, especially those under purifying or intense selection, tend to be lost quickly, and simulations may have to be repeatedly restarted.
- *Control of Disease Allele Frequency and Linkage Disequilibrium* Even if the same initial populations are used, the resulting populations will vary because of random genetic drift. The forward-time approach is directly affected by genetic drift, making it difficult to control the allele frequency at the ending generation, which makes a fair comparison of simulated samples difficult. In addition, existing implementations of forward-time simulations vary in their abilities to simulate high-density genetic markers with realistic LD patterns and none of them can readily simulate samples that use existing genetic markers in the human genome.
- *Population and Pedigree-Based Sampling* Although it is possible to collect different types of samples from a simulated population, for example, affected and unaffected individuals for a case–control design

and parents with affected offspring for an affected subpair design, existing forward-time simulation programs do not provide methods to draw samples.

This chapter focuses on simulation techniques that address these problems and demonstrates how to use forward-time simulations to simulate samples with complex human diseases.

5.2 CONTROLLING DISEASE ALLELE FREQUENCIES AT THE PRESENT GENERATION

5.2.1 Introduction of Disease Alleles

A genetic disease is caused by disease predisposing alleles at one or more loci. To simulate a genetic disease forward in time, one or more disease alleles have to be introduced to a population in some way.

The most straightforward method to introduce a disease allele to a population is to introduce it manually at a fixed generation. If the disease allele is introduced only once, all alleles at the present generation would have the same age. However, because of random genetic drift, newly introduced mutants have a small probability to survive ($1/N$ if a mutant is introduced to a population of size N). For example, if we introduce a disease allele to a population of 1000 individuals, the allele is likely to be lost after a few generations (Example 5.1).

■ EXAMPLE 5.1

This example simulates the evolution of 10 populations with 1000 individuals. These populations are initialized with random sex and all wild-type alleles (0). A `PointMutator` operator is used to introduce a disease allele (1) to the first locus of the first individual at the beginning of the evolution. The populations are evolved for 100 generations and the allele frequencies at locus 0 are calculated at the end of evolution. The last print statement prints the number of allele 1 at locus 0 in each population, using a variable `alleleNum` that is set by the `Stat` operator.

SOURCE CODE 5.1 Straightforward Simulation of the Introduction of a Disease Allele

```
>>> import simuOpt
>>> simuOpt.setOptions(quiet=True, alleleType='long')
>>> import simuPOP as sim
```

```
>>> pop = sim.Population(size=1000, loci=[1])
>>> simu = sim.Simulator(pop, 10)
>>> simu.evolve(
...     initOps=[
...         sim.InitSex(),
...         sim.PointMutator(loci=0, allele=1, inds=0)
...     ],
...     matingScheme=sim.RandomMating(),
...     finalOps=sim.Stat(alleleFreq=0),
...     gen = 100
... )
(100, 100, 100, 100, 100, 100, 100, 100, 100, 100)
>>> print([x.dvars().alleleNum[0][1] for x in simu.populations()])
[0, 0, 0, 0, 0, 0, 0, 0, 0, 0]
```

Because disease alleles introduced at a fixed generation will most likely get lost, it is natural to reintroduce the disease allele when it is lost and terminate the evolution when a certain disease allele frequency is reached. For example, a disease allele can be reintroduced if it gets lost in the population, and the evolution process stops only after the disease has been established in the population with an appreciable frequency (Example 5.2). Although this simulation method has the advantage that all populations have the same disease allele frequency, multiple attempts have to be made to introduce the disease allele, which results in different lengths of evolution time. Because many population properties such as the level of linkage disequilibrium depend on evolution time, this method is inappropriate for the simulation of disease loci with surrounding markers. The following section tries to address this problem by simulating trajectories of disease allele frequencies and evolving populations that condition on these trajectories.

■ EXAMPLE 5.2

This example simulates an evolutionary process where a disease allele is introduced whenever there is no disease allele in the population (`alleleFreq[0][1]=0`), and the simulation terminates when the disease allele reaches a frequency of 0.05. This is achieved by a `Stat` operator that calculates the allele frequency at the beginning of each generation, an `IfElse` operator that calls a `PointMutator` to introduce a disease allele whenever the allele frequency is zero, and a `TerminateIf` operator that terminates the evolution when the frequency of the disease allele is greater than or equal to 0.05.

Because disease alleles are easily lost, many attempts will be needed before an allele survives and reaches a sizable frequency. As a matter of fact, among the five populations simulated in Example 5.2, the population with the shortest evolution length needed 28 attempts and 232 generations, and the population that evolved longest needed 352 attempts and 3047 generations. This example uses a list `introGen` to store the generation number at which a mutant is introduced, using two `PyExec` operators that execute two Python statements.

SOURCE CODE 5.2 Reintroduction of a Disease Allele When It Is Lost

```
>>> import simuOpt
>>> simuOpt.setOptions(quiet=True, alleleType='long')
>>> import simuPOP as sim
>>> pop = sim.Population(size=1000, loci=[1])
>>> simu = sim.Simulator(pop, 5)
>>> simu.evolve(
...     initOps=[
...         sim.InitSex(),
...         sim.PyExec('introGen=[]')
...     ],
...     preOps=[
...         sim.Stat(alleleFreq=0),
...         sim.IfElse('alleleFreq[0][1] == 0', ifOps=[
...             sim.PointMutator(loci=0, allele=1, inds=0),
...             sim.PyExec('introGen.append(gen)')
...         ]),
...         sim.TerminateIf('alleleFreq[0][1] >= 0.05')
...     ],
...     matingScheme=sim.RandomMating()
... )
(1963, 773, 613, 232, 417)
>>> # number of attempts
>>> print([len(x.dvars().introGen) for x in simu.populations()])
[271, 70, 35, 28, 54]
>>> # age of mutant
>>> print([x.dvars().gen - x.dvars().introGen[-1] for x in simu.populations()])
[55, 73, 196, 133, 70]
```

5.2.2 Trajectory of Disease Allele Frequency

The idea of trajectory simulation has been used by others [10, 11], in the context of coalescent simulations. For example, Wang and Rannala [11] used an additive selection model and a forward approach with a normal approximation to the binomial selection process. This method can handle one disease predisposing locus and arbitrary demographic models. Coop and Griffiths [10] used diffusion approximation and a backward approach to simulate the trajectory of the allele frequency of a single locus in a population with a constant size. Our method extends these methods and also the method described in Slatkin [12].

We assume that the population size at generation t is N_t. The locus discussed is diallelic with wild-type allele A and disease allele a. Relative fitnesses of genotypes AA, Aa, and aa are 1, $1 + s_1$, and $1 + s_2$, respectively. s_1 and s_2 can assume any value greater than -1. Allele a is called advantageous if $s_i > 0$, and deleterious if $s_i < 0$ ($i = 1, 2$). s_1 and s_2 can take different signs, as in the case of balanced selection.

Suppose that disease allele a is introduced to a population at generation 1 and spreads according to a Wright–Fisher model with varying population size and a selection model described above. At generation T, the population is surveyed and i copies of allele a are found. We are interested in simulating the trajectory $H = \{i_0 = 0,\ i_1 = 1, ..., i_T = i\}$, where i_t is the number of copies of allele a at generation t. The length T of the trajectory is the age of the mutant.

The dynamics of allele frequency x_t can be modeled as follows: Assume that at generation $t - 1$, there are i_{t-1} copies of allele a. Population allele frequency is equal to $x_{t-1} = \frac{i_{t-1}}{2N_{t-1}}$. Assume that the next generation is formed from an infinite-sized gene pool. The expected frequency of allele a at generation t is expressed by the following [12]:

$$x'_t = \frac{(1 + s_1)\, x_{t-1}\,(1 - x_{t-1}) + (1 + s_2)\, x_{t-1}^2}{(1 - x_{t-1})^2 + 2\,(1 + s_1)\, x_{t-1}\,(1 - x_{t-1}) + (1 + s_2)\, x_{t-1}^2}, \tag{5.1}$$

$$= x_{t-1} \frac{1 + s_2 x_{t-1} + s_1\,(1 - x_{t-1})}{1 + s_2 x_{t-1}^2 + 2 s_1 x_{t-1}\,(1 - x_{t-1})}. \tag{5.2}$$

Therefore, the probability that there are i_t copies of allele a at generation t, given the population size N_t, equals

$$\Pr(i_t \mid i_{t-1}) = \binom{2N_t}{i_t} x_t'^{\,i_t} \left(1 - x'_t\right)^{2N_t - i_t}, \tag{5.3}$$

where x'_t is the expected allele frequency as opposed to the real allele frequency $x_t = \frac{i_t}{2N_t}$.

5.2.3 Forward- and Backward-Time Simulations

Equations 5.1 and 5.3 provide a basis for a rejection-sampling algorithm to simulate allele frequency trajectories in a forward-time manner. One may start from a single disease mutant (or multiple mutants if starting from an existing allele frequency) and simulate allele frequencies at each generation until generation T. The resulting trajectory will be accepted if x_T is within a designed range or rejected otherwise.

This algorithm works in principle and is used by programs such as GeneArtisan [11]. However, it suffers from several major problems:

- If T is large, the disease allele is under strong purifying selection, or if the acceptance region is too narrow, the acceptance probability of a trajectory is small. Obtaining one valid trajectory may require millions of attempts.
- This method assumes a fixed T (age of mutant), but T should be random. Unbiased samples of the trajectories can only be simulated if T is chosen randomly from its distribution, which is usually unknown. If an inappropriate T is chosen, the simulated trajectories will be biased.

An alternative to the forward-time algorithm is a backward approach, which was first explored by Slatkin [12] in a monogenic disease setting. Using this approach, a trajectory can be generated by a model that assumes i copies of allele a at $t = T$ and proceeds backward in time until the allele is lost. The generation at which the allele is lost becomes generation 0, if there is exactly one copy of allele a at generation 1. This approach avoids the mentioned problems of the forward-time approach.

The basic idea is to match the forward process as close as possible by reversing the Equation 5.3 with an appropriately inverted Equation 5.1. This can be achieved by solving the equation

$$x_t = x'_{t-1} \frac{1 + s_2 x'_{t-1} + s_1 \left(1 - x'_{t-1}\right)}{1 + s_2 x'^2_{t-1} + 2 s_1 x'_{t-1} \left(1 - x'_{t-1}\right)} \tag{5.4}$$

for x'_{t-1}, with known i_t, x_t, N_t, and N_{t-1}. Equation 5.4 is obtained from (5.1) by replacing x_{t-1} (now unknown) by x'_{t-1} and replacing x'_t (now known) by its sample value x_t. Because Equation 5.4 always has a unique solution (between 0 and 1), the backward binomial selection can be done for all combinations of s_1, s_2, and x_t. Given N_{t-1} and x'_{t-1}, we can then simulate i_{t-1} by

$$\Pr\left(i_{t-1} \mid i_t\right) = \binom{2N_{t-1}}{i_{t-1}} x'^{i_t}_{t-1} \left(1 - x'_{t-1}\right)^{2N_{t-1} - i_{t-1}}. \tag{5.5}$$

Note that this process can naturally simulate varying population size and selection pressure because Equation 5.5 concerns only the size and selection coefficients of the previous generation.

■ EXAMPLE 5.3

simuPOP provides functions `simulateForwardTrajectory` and `simulateBackwardTrajectory` in a utility module `simuPOP.utils` to simulate allele frequency trajectories forward- or backward-in-time. A forward-time trajectory simulation starts from a specified generation with specified allele frequencies at one or more loci. The simulator simulates allele frequencies forward in time, until it reaches a specified ending generation. In contrast, a backward-time trajectory simulation starts from the ending generation with specified allele frequencies and simulates backward in time until the allele gets lost. Both of these functions handle more unlinked loci or more arbitrary demographic models and varying fitness values. A `Trajectory` object will be returned if a trajectory is simulated successfully. This object provides a number of functions to determine the length and other properties of a trajectory. For example, a member function `freq(gen, subPop)` can be used to retrieve the frequency of all loci at specified generation and subpopulation.

This example simulates the trajectories of several alleles under purifying, balancing, or varying selection pressures under a constant or exponential population expansion model. In the first example, a locus with an initial frequency of 0.2 evolves for 500 generations and ends at a frequency between 0.1 and 0.11. A constant population size of 4000 individuals is used. The second example uses a different demographic model where a population of size 4000 expands to a size of 10,873 in 100 generations. Because genetic drift is weaker in larger populations, the allele frequency trajectories in Figure 5.1b are smoother than those in Figure 5.1a. The last two examples simulate the trajectories of disease predisposing alleles under balancing and varying selection pressures. Because the allele is under positive purifying selection in the last example, the allele frequencies in Figure 5.1d rise to higher frequencies before they return to an ending allele frequency between 15% and 16%.

SOURCE CODE 5.3 Simulating Allele Frequency Trajectory Forward in Time

```
import math
from simuPOP.utils import *
# case 1: constant population size
traj = simulateForwardTrajectory(N=4000, fitness=[1, 0.999, 0.998],
    beginGen=0, endGen=500, beginFreq=0.2, endFreq=[0.1, 0.11])
# case 2: exponential population expansion
def Nt(gen):
    return int(4000*math.exp(gen*0.01))

traj = simulateForwardTrajectory(N=Nt, fitness=[1, 0.999, 0.998],
    beginGen=0, endGen=500, beginFreq=0.2, endFreq=[0.1, 0.11])
# case 3: balancing selection
traj = simulateForwardTrajectory(N=4000, fitness=[1, 1.001, 0.998],
    beginGen=0, endGen=500, beginFreq=0.2, endFreq=[0.2, 0.22])
# case 4: varying selection pressure
def fitnessFunc(gen, subPop):
```

```
    if gen > 200:
        return (1, 0.996, 0.994)
    else:
        return (1, 1, 1.02)

traj = simulateForwardTrajectory(N=4000, fitness=fitnessFunc,
    beginGen=0, endGen=500, beginFreq=0.2, endFreq=[0.15, 0.16])
```

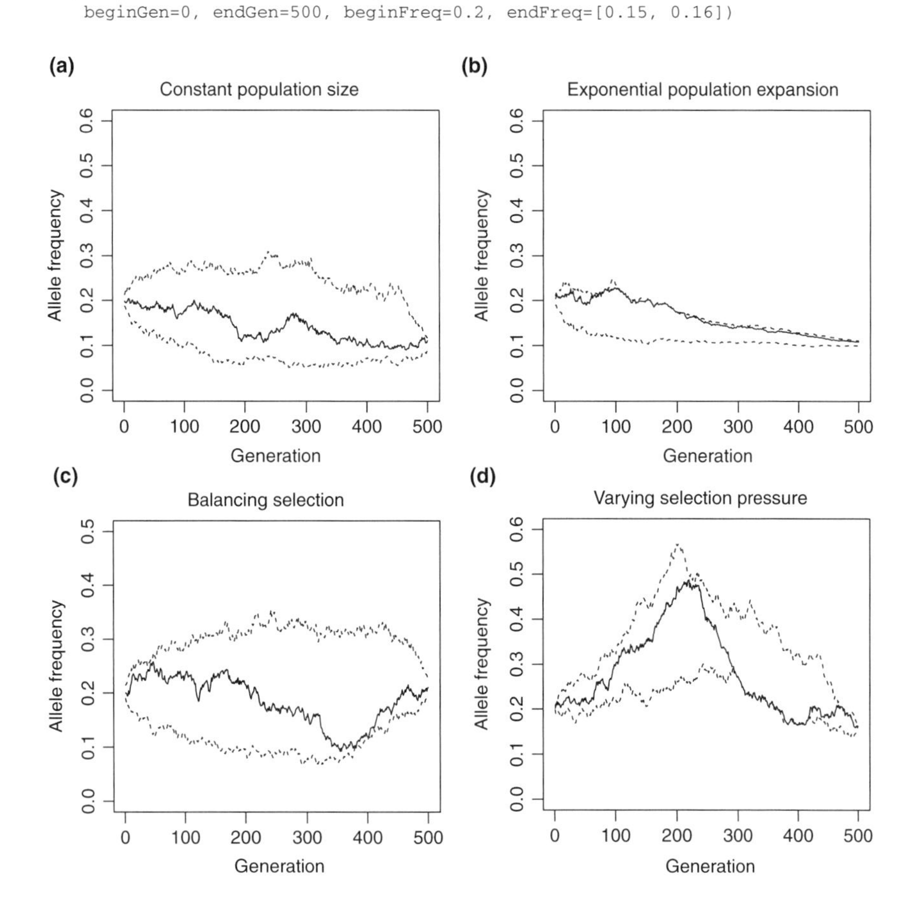

FIGURE 5.1 Examples of forward-time simulated allele frequency trajectories. Sample trajectories and minimal and maximum allele frequencies of 100 simulated trajectories of four disease predisposing alleles. A constant population size of 4000 individuals (a, c, and d) and an exponential population expansion model (b) are used. The alleles are under a purifying selection model with $s = -0.001$ (a and b), a balancing selection model with fitness values 1, 1.001, and 0.998 for genotypes AA, Aa and aa, respectively (c), and a varying selection pressure where a positive recessive selection model with $s = 0.02$ is used for the first 200 generations and an additive purifying selection model with $s = -0.003$ is used for the rest of the generations. Dotted lines in each figure represent minimal and maximum allele frequencies at each generation for 100 simulated trajectories.

Backward trajectory simulations are more interesting because simulated trajectories can have different lengths. If an allele is under positive selection, it can reach a sizable frequency in a relatively short period of time, and the simulation will succeed most of the time. If an allele is under purifying selection, it can be difficult to simulate a trajectory because most of the trajectories will become fixed instead of lost (allele frequency will increase if we look backward in time), and only the latter yield a successful trajectory. If an allele is neutral, the success rate will depend on the present allele frequency and on whether or not there is a limit on the minimum or maximum trajectory length.

Due to the constant downward drift of allele frequency, it is difficult for alleles under purifying selection to reach relatively high frequencies. Because dramatic changes of allele frequencies are more likely to happen in small populations, reaching a sizable frequency during a bottleneck is a common way for an allele under purifying selection to become common in the present population.

However, if an allele under purifying selection has reached a certain present allele frequency, the length of its trajectory does not have to be longer than an allele under positive selection. As a matter of fact, the age of a mutant is invariant to the sign of selection coefficient [13]. This result is counterintuitive but it reflects the fact that an allele under purifying selection has to reach a high allele frequency fast before it is removed from the population.

■ EXAMPLE 5.4

This example simulates trajectories of alleles under different selection and demographic models. The first four examples (Figure 5.2a–d) use a constant population size of 50,000 and the last two examples (Figure 5.2e and f) use an exponential population growth model where the population size is constant at $N_0 = 10^4$ 5000 years ago and then grows exponentially to its current size 10^6. The selection models for the first demographic model are neutral ($s_1 = s_2 = 0$), advantageous ($s_1 = 0.0005, s_2 = 0.001$), deleterious ($s_1 = -0.0005, s_2 = -0.001$), and a mixed selection model in which the disease allele is advantageous before 2000 generations ago ($s_1 = 0.001$, $s_2 = 0.002$) and is under purifying selection in the recent 2000 generations ($s_1 = -0.0001, s_2 = -0.0002$). A neutral model and a model with purifying selection are used for the last two examples (Figure 5.2e and f). In all cases, the current allele frequency is 10%. For each selection model, 100 replicates are simulated and 3 trajectories corresponding to 5%, 50%, and 95% quantiles of the trajectory length are plotted.

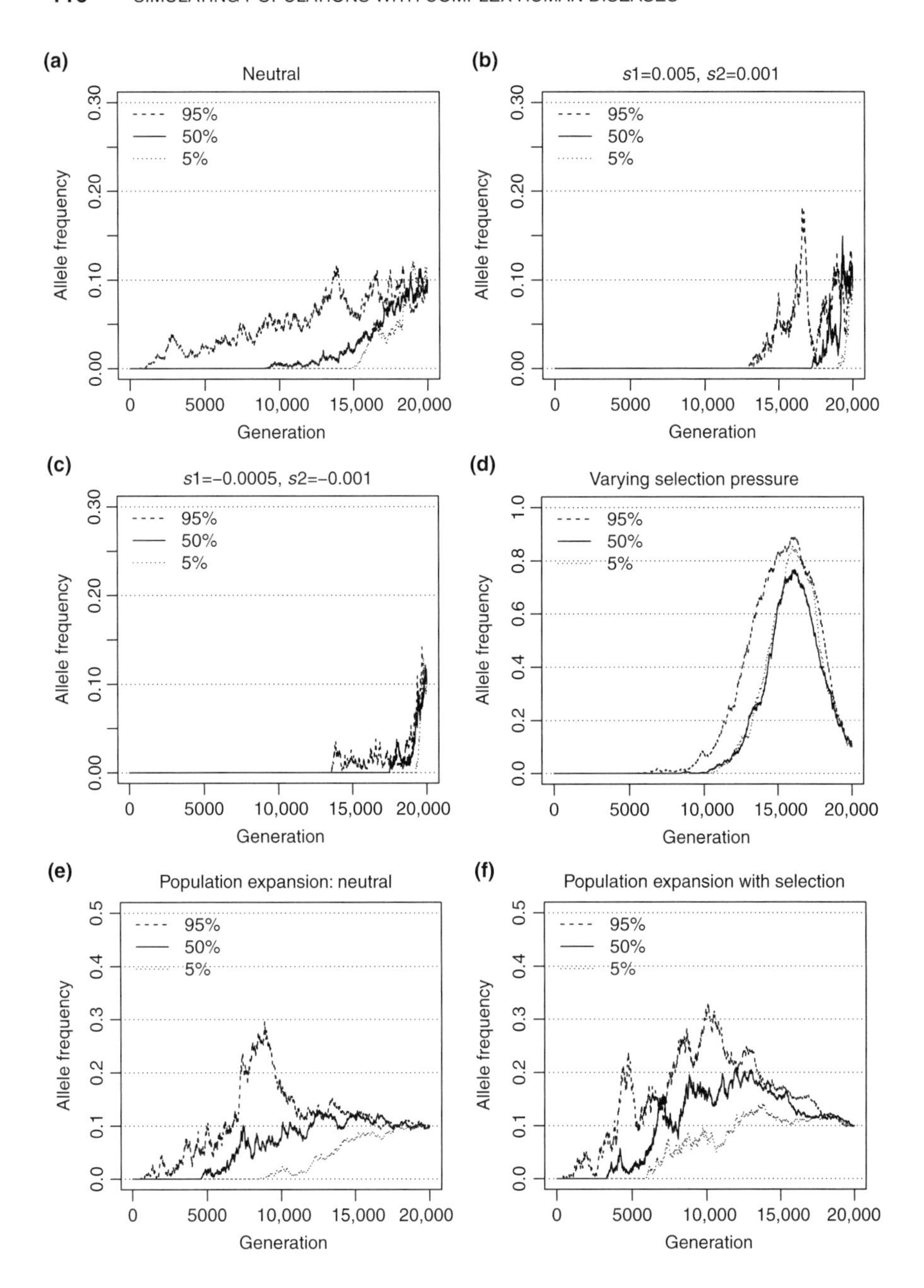

FIGURE 5.2 Examples of backward-time simulated allele frequency trajectories. Three simulated trajectories selected for 5%, 50%, and 95% quantile of trajectory lengths of 100 simulated trajectories, of four disease predisposing alleles. A constant population size of 50,000 individuals (a–d) and an exponential population expansion model (e and f) are used. The alleles are either neutral (a and e), under purifying selection models (b and f), a positive selection model (c), or varying selection pressure (d).

SOURCE CODE 5.4 Simulating Allele Frequency Trajectory Backward in Time

```
import simuPOP as sim
import math
from simuPOP.utils import *
# example 1
traj = simulateBackwardTrajectory(N=50000, fitness=[1, 1, 1],
     endGen=10000, endFreq=0.1)
# example 2
traj = simulateBackwardTrajectory(N=50000, fitness=[1, 1.001, 1.002],
     endGen=10000, endFreq=0.1)
# example 3
traj = simulateBackwardTrajectory(N=50000, fitness=[1, 0.999, 0.998],
     endGen=10000, endFreq=0.1)
# example 4
def fitnessFunc(gen, subPop):
    if gen > 8000:
        return (1, 0.999, 0.998)
    else:
        return (1, 1.001, 1.002)

traj = simulateBackwardTrajectory(N=50000, fitness=fitnessFunc,
     endGen=10000, endFreq=0.1)
# example 5
def demoFunc(gen):
    if gen < 5000:
        return 10000
    else:
        return int(10000*math.exp(0.000921*(gen-5000)))

traj = simulateBackwardTrajectory(N=demoFunc, fitness=[1, 1, 1],
     endGen=10000, endFreq=0.1)
# example 6
traj = simulateBackwardTrajectory(N=demoFunc, fitness=[1, 0.999, 0.998],
     endGen=10000, endFreq=0.1)
```

5.2.4 Random Mating with Controlled Disease Allele Frequency

With simulated allele frequency trajectories of disease predisposing loci, it is necessary to develop a method to perform random mating while controlling the disease allele frequency during evolution. The rejection-sampling algorithm, in which the next generation is regenerated if its allele frequency does not match the simulated one, can be used in principle. However, this algorithm is not efficient for practical use, especially when more than one disease predisposing locus are involved.

Controlled random mating has been used in the framework of coalescence in case of haploid populations [14, 15]. The algorithm separates generation $t-1$ and t into case and control groups and generates offspring of the case and control groups at generation t from their counterparts at generation $t - 1$.

The above works for a haploid population with one disease predisposing locus because of the independent segregation of wild-type and disease alleles. However, it does not work for a diploid population in which the wild-type and disease alleles cosegregate as heterozygotes. For a diploid population, an approximate algorithm can be used. This algorithm splits the random mating process into two stages: (i) A rejection-sampling method is applied so that only individuals with disease alleles are accepted until we obtain enough disease alleles to fit the simulated frequency trajectory. (ii) Only individuals with no disease allele are accepted; they fill the rest of the offspring generation.

Cosegregation of multiple loci because of selection against multiple disease predisposing loci complicates the problem even more. It is difficult, and sometimes impossible, to satisfy allele frequency requirements at all disease predisposing loci. Rather than one of several more complicated algorithms, we choose a simple extension to the diploid algorithm. During the first stage of this algorithm, we accept individuals that have any of the needed disease alleles until the frequency requirements at all disease predisposing loci are met. The second stage proceeds as usual. An obvious problem with this algorithm is that at the end of the first stage, disease alleles at some disease predisposing loci are accepted even if the allele frequency requirements at these disease predisposing loci have been met. This will result in, on the average, more disease alleles at all disease predisposing loci. The impact of this problem is generally negligible and is discussed in detail in Ref. [16].

In forward-time simulations, one mating event can produce more than one offspring. Because the relationship between offspring of the same family is important for gene mapping methods, family structure is preserved whenever possible. In the implementation of all the algorithms described above, acceptances and rejections are family based. Namely, the whole family is accepted or rejected, depending on its contribution to the number of disease alleles.

■ EXAMPLE 5.5

simuPOP provides a mating scheme `ControlledRandomMating` to perform random mating conditioning on specified allele frequencies at one or more loci. The controlled random mating scheme accepts a user-defined trajectory function that tells the mating scheme the desired allele frequencies at each generation. Although any user-defined function that returns allele frequencies at one or more loci at each generation can be used, such a function is usually provided by the `func()` member function of a `Trajectory` object.

If a trajectory is simulated backward in time, one or more disease alleles have to bc introduced to the exact generations when the frequency of an allele jumps from zero to a positive number. Although it is clear one or more `PointMutator` should be used in this context, it can be troublesome to figure out the starting generation of a trajectory, especially when multiple disease predisposing loci in multiple subpopulations are simulated. Fortunately, the `Trajectory` object provides a function `mutants()` to return a list of mutants in the format of `(locus, generation, subPop)` and a function `mutators()` to return a list `PointMutator` that introduces mutants to the right generations. All you need to do is to add `traj.mutators()` to the premating operators of an evolutionary process. A parameter that specifies the indices of these loci is needed.

This example combines all these pieces. In this example, two alleles under positive selection pressure are introduced to a population and reach frequencies 0.1 and 0.2, respectively. An exponential population expansion model is used to expand a population from 5000 to more than 15,000 individuals in 1000 generations. According to the simulated trajectory, the first mutant is introduced to generation 168, and the second mutant is introduced to generation 577. The trajectory is then passed to a `ControlledRandomMating` scheme to control the allele frequency of loci 0 and 1. The frequencies of allele 1 at both loci are calculated at generations 500, 600, ..., 1000 and are compared with allele frequencies of the simulated trajectory. As we have discussed in the previous section, the controlled random mating scheme follows the simulated allele frequency trajectory well, except for a slight tendency to produce higher than expected frequencies at some generations.

SOURCE CODE 5.5 Using a Controlled Mating Scheme with a Backward-Simulated Trajectory

```
>>> import simuPOP as sim
>>> from simuPOP.utils import Trajectory, simulateBackwardTrajectory
>>> from math import exp
>>> def Nt(gen):
...     'An exponential population expansion model'
...     return int(5000 * exp(.00115 * gen))
...
>>> # simulate a trajectory backward in time, from generation 1000
>>> traj = simulateBackwardTrajectory(N=Nt, fitness=[1, 1.01, 1.02], nLoci=2,
...     endGen=1000, endFreq=[0.1, 0.2])
>>> # print out mutants in the format of (loc, gen, subPop)
>>> print(traj.mutants())
[(0, 168, 0), (1, 577, 0)]
```

```
>>> pop = sim.Population(size=Nt(0), loci=[1]*2)
>>> # save Trajectory function in the sim.population's local namespace
>>> # so that the sim.PyEval operator can access it.
>>> pop.dvars().traj = traj.func()
>>> pop.evolve(
...     initOps=[sim.InitSex()],
...     preOps=traj.mutators(loci=[0, 1]),
...     matingScheme=sim.ControlledRandomMating(loci=[0, 1], alleles=[1, 1],
...         subPopSize=Nt, freqFunc=traj.func()),
...     postOps=[
...         sim.Stat(alleleFreq=[0, 1], begin=500, step=100),
...         sim.PyEval(r"'%4d: %.3f (exp: %.3f), %.3f (exp: %.3f)\n' % (gen, alleleFreq[0][1],"
...             "traj(gen)[0], alleleFreq[1][1], traj(gen)[1])",
...             begin=500, step=100)
...     ],
...     gen=1001  # evolve 1001 generations to reach the end of generation 1000
... )
 500: 0.013 (exp: 0.013), 0.000 (exp: 0.000)
 600: 0.005 (exp: 0.005), 0.003 (exp: 0.003)
 700: 0.011 (exp: 0.011), 0.008 (exp: 0.008)
 800: 0.012 (exp: 0.013), 0.031 (exp: 0.031)
 900: 0.037 (exp: 0.037), 0.092 (exp: 0.092)
1000: 0.101 (exp: 0.100), 0.200 (exp: 0.200)
1001
```

5.3 FORWARD-TIME SIMULATION OF REALISTIC SAMPLES

Although simple samples simulated under idealized assumptions can be used to validate properties of statistical gene mapping methods, only samples that reflect the complex structure of the human genome and the genetic basis of human genetic diseases can be used to evaluate and compare the statistical power of these methods and to compare various sampling designs under realistic conditions. Otherwise, a gene mapping method may perform well in theory and on simulated data sets, but poorly on real data sets [17, 18].

Utilizing simulation techniques we have developed up to now, this section describes a simulation method that simulates genetic samples with realistic patterns of linkage disequilibrium. To retain the complex genetic structure of human populations, this algorithm creates an initial population of selected markers from a real sample. It then evolves this population forward in time, subject to mutation, recombination, natural selection, and rapid population expansion. This process uses an optional scaling algorithm to improve its performance when weak additive selection forces are used and uses a trajectory simulation method to control the frequency of disease predisposing alleles. Depending on specific applications, the last step of this process involves different postprocessing steps. For example, a rejection-sampling algorithm can be used to simulate case–control samples or trio families (affected offspring with parents) with rare diseases. A few examples are used to demonstrate the applications of this method.

5.3.1 Method

Because of the complexity of human genomes and their largely unknown evolutionary histories, it is infeasible to simulate samples that closely resemble human populations by evolving a simulated initial population. Therefore, our simulation method uses real empirical data sets to simulate large populations with additional genetic variations while retaining key features of the empirical data sets. Thanks to rapid advances in genotyping technology, the genotype data of millions of single-nucleotide polymorphism (SNP) markers of hundreds or even thousands of individuals are currently available [19, 20], and higher density data will become available in the near future [21]. The availability of data facilitates the creation of an initial population with selected markers that match an existing sample, which usually contains markers from commercially available genotyping platforms.

Creation of an Initial Population The first step of our simulation method is to create an initial population from a real sample with selected markers. Depending on the application, one may want to start from an existing GWA study with thousands of controls, such as the control data from the Wellcome Trust Case Control Consortium [20], or from a publicly available data set, such as phase 2 or phase 3 of the HapMap data set [19]. Our study used 993 unrelated individuals (parents in trio and duo samples and all unrelated individuals) in 10 populations of the phase 3 HapMap data set because these data are readily available. Depending on the specific application, markers can be chosen according to markers used in real-world studies (e.g., the markers on the Illumina 550k genotyping chip) or by marker distance and minor allele frequency; individuals from one or more HapMap populations can be selected either as separated populations or as a single population.

The simuPOP online cookbook provides a number of Python modules to prepare such an initial population. For example, a script `loadHapMap3.py` downloads the third phase of the HapMap data set and save them in simuPOP format and a script `selectMarkers.py` can be used to select markers and individuals according to criteria such as start and ending positions, number of markers, minor allele frequency, and minimal distance between adjacent markers, or according to a list of markers.

Demographic Model We consider the initial populations as small, isolated populations before the expansion of a typical human population (around 12,000 years or 600 generations ago, if we assume 20 years per generation with the invention of agriculture) [22]. We then expand these populations linearly to a larger population of 105 individuals by adding the same number of individuals each year, subject to mutation, recombination, and natural selection. We use linear population expansion instead of a more commonly used exponential expansion model because a linear model expands the initial population faster at first, thus better preserving genetic diversity in the initial population and resulting in a final population of a larger effective population size. For example, if we start with 993 individuals from the phase 3 HapMap sample and expand the population for 500 generations using linear and exponential population expansion models, the effective population sizes of the expanded populations with 105 individuals would be 12,658 and 4603, respectively [23]. The former is comparable to the effective population size of real human populations.

Evolving the Founder Population During evolution, we mutate all SNP markers according to a symmetric diallelic mutation model with a mutation rate of 10^{-8} per base pair per generation. At each generation, parents are chosen at random and pass their genotypes to offspring according to Mendelian laws. Parental chromosomes are recombined according to a fine-scale genetic map estimated from the HapMap data set [24] before one of the recombinants is passed to an offspring. If a selection model is specified, parents are chosen with probabilities that are proportional to their relative fitness values. Our simulation method supports both single-locus and multilocus natural selection models, including models that involve multiple interacting DPL. If multiple populations are simulated, a stepping-stone migration model with a low migration rate is applied to control the genetic distance between the populations [25].

A scaling approach is used to improve the efficiency of our simulation [26]. Compared to a regular simulation that evolves a population of size N for t generations, a scaled simulation with a scaling factor λ evolves a smaller population of size N/λ for t/λ generations with magnified (multiplied by λ) mutation, recombination, and selection forces. This method could be justified by a diffusion approximation to the standard Wright–Fisher process [23, 26]; however, because the diffusion approximation applies only to weak genetic forces in the evolution of haploid sequences, it cannot be used when nonadditive diploid or strong genetic forces are used. Our simulation program simulates populations with specified population size, so a population simulated using a scaling factor λ would be comparable to an unscaled simulation of a population that is λ times larger.

Control of Disease Allele Frequency To simulate a genetic disease, we control the frequencies of DPAs at DPL using presimulated allele frequency trajectories. Either a forward-time approach or a backward-time approach can be applied. More specifically, if we assume that a DPA existed before population expansion, we simulate the frequency of DPA forward in time until it reaches the present generation. The simulation starts from the frequency of DPA in the initial population and is restarted if the allele frequency at the present generation falls out of the desired range. If the mutant is recent (e.g., appears within the past 500 generations), we simulate from the frequency of DPA at the present generation backward in time until the allele gets lost. Multilocus natural selection models are supported with the restriction that DPAs have to be unlinked. After the allele frequency trajectories of DPAs are simulated, we use a special random mating scheme to evolve the population forward in time while following the simulated trajectories at these loci.

Sample Generation The final postprocessing step of the simulation process will vary depending on the individual application. To simulate a common disease with enough affected individuals in the simulated population, we can draw samples directly from the population after the affection status of each individual is determined, usually using a penetrance model that yields the probability that an individual is affected with a disease according to his or her genotype (Pr(affection status | genotype)). Alternatively, a rejection-sampling algorithm could be used to draw case–control samples or samples with independent offspring (such as trios) of a rare disease. More specifically, we choose parents from the simulated population and produce offspring repeatedly, apply the penetrance model to determine the affection status of each offspring, and continue the process until enough samples are collected.

■ EXAMPLE 5.6

Example 5.6 implements this simulation method and provides a function `simuGWAS` that will be used to simulate data for examples in the rest of this chapter. This script defines a function `linearExpansion` (lines 5–14 in Source Code 5.6) that returns a demographic function from specified starting and ending population size and number of generations to evolve, and then a function `simuGWAS` that evolves a population forward

TABLE 5.1 Parameters of Function SimuGWAS in Example 5.6

Parameter	Usage
`pop`	A starting population saved in simuPOP format
`mutaRate`	Mutation rate for a dialleleic mutation model with the same forward and backward mutation rates, default to 10^{-8}
`recIntensity`	Recombination rate per base pair per generation, default to 10^{-8}
`migrRate`	Migration rate
`expandGen`	Number of generations to evolve
`expandSize`	Final population size, it can be an overall population size or a list of subpopulation sizes if the initial population has multiple subpopulations
`DPL`	Names of disease predisposing loci
`curFreq`	Frequency of allele 1 at the present generation
`fitness`	A list of fitness values. Should be a length of three times number of DPL
`scale`	A scaling parameter. This will effectively boost mutation, migration, and recombination rate and reduce final population size
`logger`	A Python logging object. If specified, it will be used to output progress and debug information depending on the level of logging

in time, subject to mutation, recombination, migration (if there are multiple subpopulations), population expansion, and natural selection.

Function `simuGWAS` accepts a number of parameters that specify the demographic and genetic models used for the evolutionary process (Table 5.1). Lines 20–24 implement the scaling technique, which merely multiplies or divides parameters with a scaling factor. If a list of disease predisposing loci is specified, the frequencies of allele 1 at these loci are checked. If there is no existing allele at these loci, a backward allele frequency trajectory simulation algorithm is used to introduce a mutant between generation 0 and `expandGen`. Otherwise, a forward-time trajectory simulation algorithm is used to evolve an existing allele to expected allele frequency. For the sake of simplicity, this function does not handle mixed forward and backward trajectories.

The rest of the function uses the `Population.evolve` function to evolve the population for `expandGen` generations using a controlled random mating scheme. This function outputs the generation number and population size at the end of each generation, and a measure of population structure (F_{ST}) at every 10 generations if there are multiple subpopulations.

SOURCE CODE 5.6 Simulation of Populations with Realistic Pattern of Linkage Disequilibrium

```
import simuPOP as sim
from simuPOP.utils import simulateForwardTrajectory, simulateBackwardTrajectory, \
    migrSteppingStoneRates

def linearExpansion(N0, N1, G):
    '''Return a linear population expansion demographic function that expands
    a population from size N0 to N1 linearly in G generations. N0 and N1 should
    be a list of subpopulation sizes.'''
    step = [float(x-y) / G for x,y in zip(N1, N0)]
    def func(gen):
        if gen == G - 1:
            return N1
        return [int(x + (gen + 1) * y) for x, y in zip(N0, step)]
    return func

def simuGWAS(pop, mutaRate=1.8e-8, recIntensity=1e-8, migrRate=0.0001,
    expandGen=500, expandSize=[10000], DPL=[], curFreq=[], fitness=[1,1,1],
    scale=1, logger=None):
    # handling scaling...
    mutaRate *= scale
    recIntensity *= scale
    migrRate *= scale
    expandGen = int(expandGen / scale)
    fitness = [1 + (x-1) * scale for x in fitness]
    pop.dvars().scale = scale
    # Demographic function
    demoFunc = linearExpansion(pop.subPopSizes(), expandSize, expandGen)
    # define a trajectory function
    trajFunc = None
    introOps = []
    if len(DPL) > 0:
        stat(pop, alleleFreq=DPL, vars='alleleFreq_sp')
        currentFreq = []
        for sp in range(pop.numSubPop()):
            for loc in pop.lociByNames(DPL):
                currentFreq.append(pop.dvars(sp).alleleFreq[loc][1])
        # if there is no existing mutants at DPL
        if sum(currentFreq) == 0.:
            endFreq=[(x-min(0.01,x/5.), x+min(0.01, x/5., (1-x)/5.)) for x in curFreq]
            traj=simulateForwardTrajectory(N=demoFunc, beginGen=0, endGen=expandGen,
                beginFreq=currentFreq, endFreq=endFreq, nLoci=len(DPL),
                fitness=fitness, maxAttempts=1000, logger=logger)
            introOps=[]
        else:
            traj=simulateBackwardTrajectory(N=demoFunc, endGen=expandGen, endFreq=curFreq,
                nLoci=len(DPL), fitness=fitness, minMutAge=1, maxMutAge=expandGen,
                logger=logger)
            introOps = traj.mutators(loci=DPL)
        if traj is None:
            raise SystemError('Failed to generated trajectory after 1000 attempts.')
        trajFunc=traj.func()
    if pop.numSubPop() > 1:
        pop.addInfoFields('migrate_to')
    pop.dvars().scale = scale
```

```
pop.evolve(
    initOps=sim.InitSex(),
    preOps=[
        sim.SNPMutator(u=mutaRate, v=mutaRate),
        sim.IfElse(pop.numSubPop() > 1,
            sim.Migrator(rate=migrSteppingStoneRates(migrRate, pop.numSubPop()))),
        ] + introOps,
    matingScheme=sim.ControlledRandomMating(loci=DPL, alleles=[1]*len(DPL),
        freqFunc=trajFunc, ops=sim.Recombinator(intensity=recIntensity),
        subPopSize=demoFunc),
    postOps = [
        sim.Stat(popSize = True, structure=range(pop.totNumLoci())),
        sim.PyEval(r'"After %3d generations, size=%s\n" % ((gen + 1 )* scale, subPopSize)'),
        sim.IfElse(pop.numSubPop() > 1,
            sim.PyEval(r"'F_st = %.3f\n' % F_st", step=10), step=10),
    ],
    gen = expandGen
)
return pop
```

5.3.2 Drawing Population and Family-Based Samples

In real-world genetic studies, we never have the luxury to collect genotype and phenotype information for all individuals in a population and have to base our analysis on collected samples. To evaluate the performance of statistical gene mapping methods using simulated data sets, it is necessary to draw samples from simulated populations.

To simulate a genetic disease, a penetrance model is needed to assign affection status of individuals in a population according to his or her genotype and other properties. For example, using a penetrance table reported in Ford et al., the probabilities that a BRCA1 carrier gets breast cancer are 0.036, 0.18, 0.57, 0.75, and 0.83 when she is 30, 40, 50, 60, and 70 years of age, respectively [27]. Example 5.7 demonstrates how to define a penetrance function and apply it to a population in simuPOP.

■ EXAMPLE 5.7

This example implements a disease model of breast cancer according to a penetrance model [27]. Because single penetrance operators such as `MapPenetrance` only consider an individual genotype, we use a `PyPenetrance` operator that calculates individual penetrance according to a user-provided callback function. This callback function takes individual genotype (parameter `geno`) and information fields as parameters and returns a penetrance probability.

This example creates a population of 10,000 individuals with a single-locus BRCA1 with the frequency of the rare allele being an unrealistically high value of 0.27. Each individual is assigned a random age between

0 and 70. After a penetrance model is applied, a `Stat` operator is used to calculate the population prevalence of this disease. Although the example uses function forms of all operators, it would be easy to assign individual fitness and track the prevalence of this disease dynamically during an evolutionary process. This example also demonstrates how to define virtual subpopulations by age and genotype (carrier or not) and how to calculate statistics for individuals within each virtual subpopulation.

SOURCE CODE 5.7 A Single-Locus Penetrance Model of Breast Cancer

```
>>> import simuPOP as sim
>>> import random
>>>
>>> def breatCancer(geno, age):
...     # brca1 is recessive
...     if 0 in geno:
...         return 0
...     return [0, 0, 0.036, 0.18, 0.57, 0.75, 0.83][int(age/10)]
...
>>> pop = sim.Population(size=10000, loci=1, infoFields='age',
...     lociNames='BRCA1')
>>> sim.initGenotype(pop, freq=[0.73, 0.27])
>>> sim.initInfo(pop, lambda: random.randint(0, 69), infoFields='age')
>>> sim.pyPenetrance(pop, func=breatCancer, loci=0)
>>> pop.setVirtualSplitter(sim.CombinedSplitter([
...     sim.InfoSplitter(field='age', cutoff=[30,40,50,60]),
...     sim.ProductSplitter([
...         sim.InfoSplitter(field='age', cutoff=[30,40,50,60]),
...         sim.GenotypeSplitter(loci=0, alleles=[1,1], names='carrier'),
...     ])]
... ))
>>> sim.stat(pop, numOfAffected=True, subPops=[(0, sim.ALL_AVAIL)],
...     vars=['propOfAffected', 'propOfAffected_sp'])
>>> print('Population prevalence is %.2f%%' % (pop.dvars().propOfAffected*100))
Population prevalence is 4.56%
>>> for x in range(pop.numVirtualSubPop()):
...     print('Prevalence in group %s is %.2f%%' % \
...         (pop.subPopName((0,x)), pop.dvars((0,x)).propOfAffected*100))
...
Prevalence in group age < 30 is 0.07%
Prevalence in group 30 <= age < 40 is 1.22%
Prevalence in group 40 <= age < 50 is 3.69%
Prevalence in group 50 <= age < 60 is 6.06%
Prevalence in group age >= 60 is 6.12%
Prevalence in group age < 30, carrier is 0.95%
Prevalence in group 30 <= age < 40, carrier is 18.48%
Prevalence in group 40 <= age < 50, carrier is 58.70%
Prevalence in group 50 <= age < 60, carrier is 77.78%
Prevalence in group age >= 60, carrier is 86.87%
```

A case–control design is a type of retrospective study design that has been widely used in epidemiological studies, especially genome-wide association studies. In a case–control study, people with a specific disease (cases) are chosen with people who do not have the disease (controls). The basic idea is to compare genotypes of cases and controls. If alleles or genotypes at a locus is significantly different in cases and controls, these alleles or genotypes are claimed to be associated with the disease status. Because disease outcome might be influenced by other characteristics such as sex and race, cases and controls are matched so that these characteristics are similar in these two groups.

If a disease is common enough so that there are enough cases in a simulated population, it is easy to draw case–control samples from a population. For example, Example 5.8 simulates a disease that is caused by a single locus with penetrance 0.10, 0.12, and 0.20 for genotypes 00, 01, and 11, respectively. With a disease allele frequency of 30%, the expected disease prevalence is 11.74%, which results in more than 1000 affected individuals in a population of 10,000 individuals. This example draws 500 cases and 500 controls from the population and applies allele-based χ^2 tests to assess the association between disease status and genotype at each locus. As we can see from the p-values of these tests, the disease-causing locus (locus 2) is strongly associated with the disease. Because no linkage disequilibrium exists between these simulated loci, no association is detected for loci around the disease-causing locus.

■ EXAMPLE 5.8

This example creates a population of 10,000 individuals, each with a chromosome with five loci. Function `initGenotype`, which is the function form of operator `InitGenotype`, is used to initialize the population with a frequency of 0.3 for allele 1. A penetrance operator is then used to set the affection status of each individual. A function `drawCaseControlSample` defined in a utility module `simuPOP.sampling` is then used to draw a sample of 500 cases and 500 controls. The sample is analyzed using a `Stat` operator that performs an allele-based χ^2 tests to assess the association between disease status and genotype at each locus. The p-values of these tests are stored in variable `Allele_ChiSq_p` and are printed by the last print statement.

SOURCE CODE 5.8 Draw Case–Control Samples

```
>>> import simuPOP as sim
>>>
>>> from simuPOP.sampling import drawCaseControlSample
>>> # create a population with affected individuals
>>> pop = sim.Population(size=10000, loci=5)
```

```
>>> sim.initGenotype(pop, freq=[0.7, 0.3])
>>> sim.maPenetrance(pop, loci=2, penetrance=[0.1, 0.12, 0.20])
>>> sim.stat(pop, numOfAffected=True, association=range(5))
>>> print(pop.dvars().numOfAffected)
1149
>>> # draw a case control sample and run association test
>>> sample = drawCaseControlSample(pop, cases=500, controls=500)
>>> sim.stat(sample, numOfAffected=True, association=range(5))
>>> print(sample.dvars().numOfAffected, sample.dvars().numOfUnaffected)
(500, 500)
>>> print(', '.join(['%.3e' % sample.dvars().Allele_ChiSq_p[x] for x in range(5)]))
1.870e-01, 8.843e-01, 9.223e-06, 3.084e-01, 7.870e-02
```

If a disease is rare or if the requested sample size is large or unknown in advance, it can be difficult to estimate the required population size to guarantee there are enough cases in the simulated population. Instead of drawing directly from a population, one solution of this problem is to draw samples from an infinitely sized offspring population. In practice, we only need to simulate an offspring population that has enough cases and we do not have to keep individuals that are not sampled. Therefore, we can produce offspring repeatedly and can keep or discard them according to their affection status, until we collect enough cases and controls.

■ EXAMPLE 5.9

This example demonstrates how to generate a sample from a population of 10,000 individuals. Because this disease has a low penetrance of only 0.01, 0.02, and 0.10 for genotypes 00, 01, and 11, respectively, the expected disease prevalence is only 1.68% with a disease allele frequency of 20%. To produce a sample of 500 cases and 500 controls from this population, we evolve it for one generation and collect cases and controls from the offspring population. This process is achieved by the use of virtual subpopulations and an operator `DiscardIf`. This example uses a `ProductSplitter` that defines four VSPs, namely, affected and unaffected offspring in the first 500 and last 500 individuals in the offspring population. By discarding all individuals that belong to virtual subpopulations (VSPs) (0, 0) and (0, 3), we discard affected offspring in the first 500 offspring and unaffected individuals in the last 500 offspring. The result of this offspring selection process is an offspring population of 500 controls followed by 500 cases.

SOURCE CODE 5.9 Generating Case–Control Samples

```
>>> import simuPOP as sim
>>> def genCaseControlSample(pop, nCase, nControl, penetrance):
...     '''Draw nCase affected and nControl unaffected individuals by producing
...     offspring from pop repeatedly until enough cases and controls are
```

```
...     collected. A penetrance operator is needed to assign affection status
...     to each offspring.
...     '''
...     sample = pop.clone()
...     sample.setVirtualSplitter(sim.ProductSplitter([
...         sim.AffectionSplitter(),
...         sim.RangeSplitter([[0, nCase], [nCase, nCase + nControl]])]))
...     sample.evolve(
...         matingScheme=sim.RandomMating(ops=[
...             sim.MendelianGenoTransmitter(),
...             penetrance,
...             sim.DiscardIf(True, subPops=[(0,0), (0,3)])],
...             subPopSize=nCase + nControl
...         ),
...         gen=1
...     )
...     return sample
...
>>> if __name__ == '__main__':
...     pop = sim.Population(size=10000, loci=1)
...     sim.initGenotype(pop, freq=[0.8, 0.2])
...     sim.initSex(pop)
...     sample = genCaseControlSample(pop, 500, 500,
...         sim.MaPenetrance(loci=0, penetrance=[0.01, 0.02, 0.10]))
...     #
...     sim.stat(sample, numOfAffected=True)
...     print(sample.dvars().numOfAffected, sample.dvars().numOfUnaffected)
...
(500, 500)
```

An affected sibpair design is an efficient study design for the mapping of rare and highly penetrant diseases. The basic idea is that siblings are expected to share alleles that are identical by descent because of their shared ancestry. If both siblings are affected, they tend to share the same disease predisposing alleles at the loci that are associated with the disease. The proportion of alleles that is identical by descent may differ from expectation based upon Medelian segregation of parental genotype.

Unless a simulated population has kept parental information during evolution, it is obviously infeasible to draw affected sibpairs and their parents from the simulated population. However, even if a population keeps parental generations during evolution, it can be difficult to draw siblings from a population if a random mating process is used to evolve it. This is because the probability that two offspring share the same parents is $1/N$, which is quite small for a reasonably sized population. Consequently, very few sibpairs could be sampled from populations simulated using a mating scheme unless more than one offspring are produced at each mating event. Similar to what we have shown for case–control samples, we can draw affected sibpair samples from a multigenerational population if there are enough affected sibpairs and their parents or generate affected sibpair samples by evolving a population for one generation. This technique is demonstrated in Example 5.10 and will be used in Example 5.12.

■ EXAMPLE 5.10

This example defines a function `genAffectedSibpairSample` that makes a clone copy of a passed population, increases its ancestral depth to keep the parental population, adds required information fields (to record parentship information), assigns unique IDs to all individuals, and then evolves the population for one generation. A function `drawAffectedSibpairSample` defined in module `simuPOP.sampling` is used to draw the required number of affected sibpair and their parents. With some more work, this function can make use of the simulation technique described in Example 5.9 and ensure that enough affected sibpairs are available in the offspring population.

SOURCE CODE 5.10 Generating Affected Sibpair Samples

```
>>> import simuPOP as sim
>>>
>>> from simuPOP.sampling import drawAffectedSibpairSample
>>>
>>> def genAffectedSibpairSample(pop, nFamilies, penetrance):
...     '''Draw nFamilies affected sibpairs and their parents by producing
...     siblings from pop repeatedly until enough affected sibpairs are
...     collected. A penetrance operator is needed to assign affection status
...     to each offspring.
...     '''
...     pop1 = pop.clone()
...     pop1.setAncestralDepth(1)
...     pop1.addInfoFields(['ind_id', 'father_id', 'mother_id'])
...     pop1.evolve(
...         initOps=sim.IdTagger(),
...         matingScheme=sim.RandomMating(
...             ops=[
...                 sim.MendelianGenoTransmitter(),
...                 penetrance,
...                 sim.IdTagger(),
...                 sim.PedigreeTagger(),
...             ],
...             numOffspring=2,
...             subPopSize=pop.popSize()*2
...         ),
...         gen=1
...     )
...     sim.stat(pop1, numOfAffected=True)
...     return drawAffectedSibpairSample(pop1, nFamilies)
...
>>> if __name__ == '__main__':
...     pop = sim.Population(size=10000, loci=1)
...     sim.initGenotype(pop, freq=[0.5, 0.5])
...     sim.initSex(pop)
```

```
...         sim.maPenetrance(pop, loci=0, penetrance=[0.05, 0.15, 0.30])
...         sample = genAffectedSibpairSample(pop, 100,
...             sim.MaPenetrance(loci=0, penetrance=[0.05, 0.15, 0.30]))
...         #
...         sim.stat(sample, numOfAffected=True)
...         print(sample.dvars().numOfAffected, sample.dvars().numOfUnaffected)
...
(200, 0)
```

5.3.3 Example 1: Typical Simulations With or Without Scaling

We created an initial population of 993 independent individuals of the HapMap phase 3 data set, using 5000 markers on region 2p16.3 (chr2:51002576-60032817). This region spans 9.03 Mbp with a genetic distance of 6.97 cM. It contains ENr112 ENCODE region and has an average marker distance of 1.81 kb. We evolved this population for 500 generations until it reached 50,000 individuals, subject to mutation (at a mutation rate of 10^{-8} per locus per generation), recombination (according to the genetic distance between adjacent markers), no selection, and linear population expansion. The initial population can be prepared using commands

```
loadHapMap3.py --gui=batch --chroms=[2]
selectMarkers.py --gui=batch --chroms=[2] --startPos=51000000 --numMarkers=5000
```

or by executing these scripts using their graphical user interfaces.

In order to evaluate the quality of simulated populations and the impact of the scaling technique, we simulated three expanded populations of sizes 50,000, 25,000, and 10,000 using scaling factors 1 (unscaled), 2, and 5, respectively, and an expanded population of 50,000 individuals using a scaling factor of 5. These can be achieved by calling the `simuGWAS` function defined in Example 5.6 with appropriate parameters.

Although the first three populations are scaled versions of the same evolutionary process, the last one is comparable to an unscaled simulation of a population of size 250,000. Compared to the first three simulations, genetic drift has a smaller impact on the last simulation because of its larger population sizes during evolution. This is demonstrated in Figure 5.3 where the allele frequencies at 5000 markers for all simulated populations are compared with those for the initial population.

The evolution of LD in such an evolutionary process is more complicated. According to Figure 5.4, all simulated populations had lower LD values than those of the initial population. Although populations simulated using a scaling approach tended to have lower LD values than those from unscaled simulations, the differences between mean R2 values were

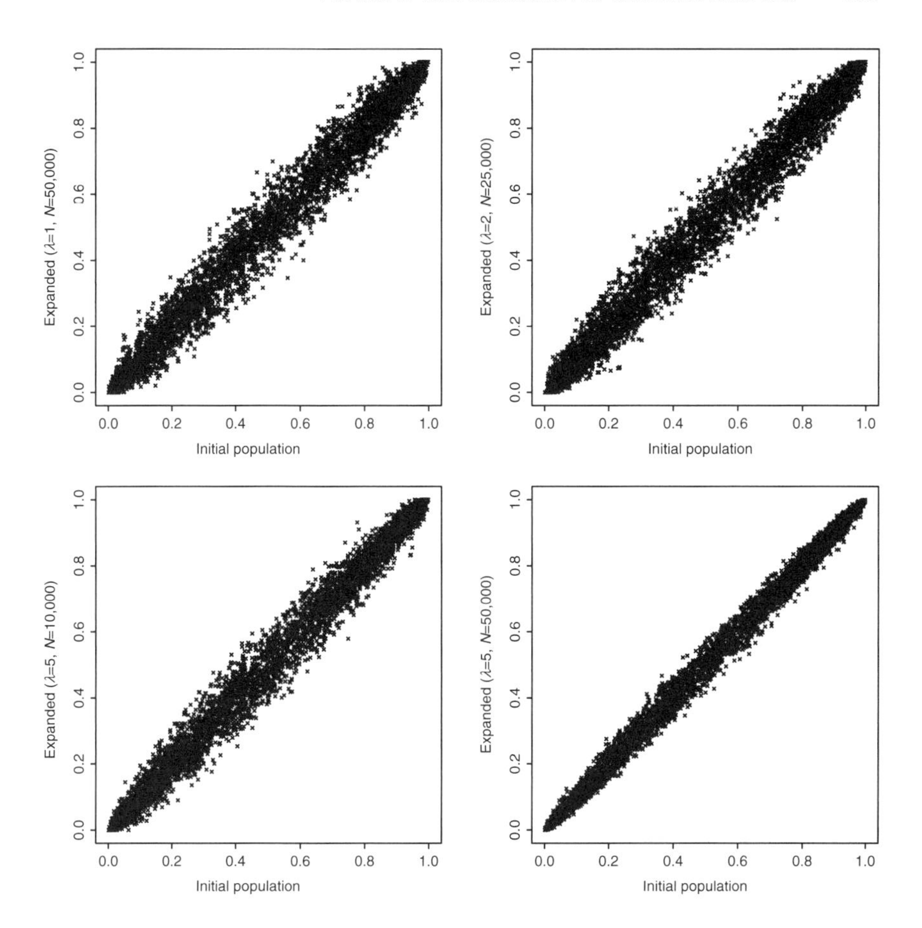

FIGURE 5.3 Comparison between allele frequencies in initial and expanded populations. Allele frequencies of the initial (x-axis) and expanded (y-axis) populations of four simulations with populations sizes 50,000, 25,000, 10,000, and 50,000 and scaling factors $\lambda = 1$ (unscaled), 2, 5, and 5, respectively.

negligible especially for markers that are less than 200 kbp apart. A more detailed analysis showed that average LD increased and then decreased during the evolutionary process of all simulations. This phenomenon could be explained by the fact that our simulation started from a relatively small population, so LD first built up because of a bottleneck effect. With increasing population size, the natural decay of LD through genetic recombination gradually prevailed at a rate accelerated by the impact of rapid population expansion [28, 29]. The simulation with a scaling factor of 5 and population size 50,000 had the lowest LD values because it had a relatively short period of bottleneck and a faster rate population expansion than other simulations.

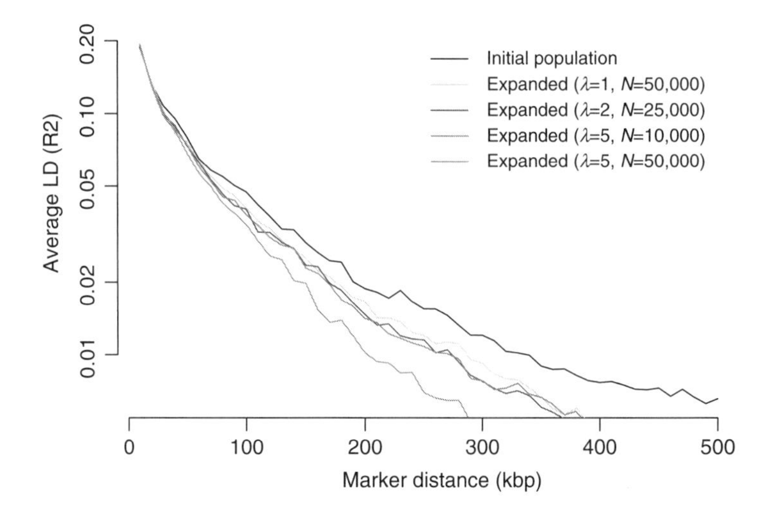

FIGURE 5.4 Decay of linkage disequilibrium as a function of marker distance. Average LD values as a function of marker distance for the initial population and four expanded populations of sizes 50,000, 25,000, 10,000, and 50,000, using scaling factors $\lambda = 1$ (unscaled), 2, 5, and 5, respectively. The y-axis is plotted in log scale to distinguish LD curves in low LD regions. Marker distances were cut into bins of 10 kbp. For example, the average LD at point 200 kbp represents the mean pairwise LD values of all pairs of markers that were 200–210 kbp apart.

5.3.4 Example 2: A Genetic Disease with Two DPL

We extracted 6000 markers (2000 markers each) on chromosomes 2 (chr2:20014298-31200250), 5 (chr5:20005983-32781509), and 10 (chr10:41756307-55305682) of 993 independent individuals from phase 3 of the HapMap sample. We selected markers from a commercially available genotyping chip (the Illumina 550k array) to match markers used in real-world GWAS. The average distance between adjacent markers was 5.61 kb.

Two markers, rs4491689 (chr2:26494285) and rs6869003 (chr5:27397573), were selected to be DPL of a genetic disease. The first marker was assumed to be under purifying selection, with fitness values of 1, 0.996, and 0.994 for genotypes *AA*, *AG*, and *GG*, respectively. The second marker was assumed to be under positive selection, with fitness values of 1, 1.001, and 1.005 for genotypes *CC*, *CT*, and *TT*, respectively. A multiplicative multilocus selection model was used. Because these two loci reside on different chromosomes, we considered natural selection to be applied to these loci independently [30]. We assumed that DPAs existed longer than 500 generations, and we used a forward-time simulation method to simulate trajectories of the frequencies of the minor alleles at both loci, starting from their frequencies in the chosen HapMap sample (0.28 for marker rs4491689 and 0.07 for marker rs6869003). The ending allele frequencies were 0.05 for marker rs4491689 and 0.15 for marker rs6869003, which were chosen in concordance with the selection pressurc that was applied to each marker. We did not scale this simulation because of the use of a nonadditive diploid selection model.

We used a logistic model with gene–gene and gene–environment interactions to model a disease that involves these two genetic markers and a random environmental factor with two states 0 and 1. We assumed that the disease was mild and was not the source of the selection pressure on the two DPL. The model can be expressed as

$$\begin{aligned} Y = \text{logit}\,(\Pr(Y = 1 \mid g_1, g_2, e)) &= \alpha + \beta_1 g_1 + \beta_2 g_2 + \beta_3 g_1 g_2 \\ &\quad + \gamma_1 g_1 e + \gamma_2 g_2 e, \end{aligned} \tag{5.6}$$

where Y is the disease status, g_1 and g_2 are number of DPAs at two markers, respectively, and e is the random environmental factor. This model is an extension of the one-gene, one-environmental model used in Li and Conti [31]. We chose positive α, β_i, and γ_i values so that the presence of each DPA increases the probability that an individual is affected with the disease. We chose $\beta_1 = \beta_2/2$, $\gamma_1 = \gamma_2/2$ so that DPA at the marker rs4491689 had less impact on the disease than DPA at marker rs6869003. Finally, we controlled parameters α, β_i, and γ_i so that the prevalence of the disease was 1%. Because there were less than 1000 affected individuals in the expanded population, we used a rejection-sampling algorithm, as demonstrated in Example 5.9, to populate an offspring population of exactly 1000 cases and 1000 controls from the expanded population.

■ EXAMPLE 5.11

We use scripts `loadHapMap3.py` and `selectMarkers.py` to prepare an initial population. In order to select markers from the Illumina array, we extracted marker names from the array annotation file and passed them to script `selectMarkers.py`. The population is evolved using function `simuGWAS` defined in Example 4.6.

Because penetrance model defined by Equation 5.6 requires several parameters, this example defines a function `penetrance` that returns a penetrance function from parameters α, β_i, and γ_i. The returned function is passed to function `genCaseControlSample` defined in Example 5.9 to produce a sample of 1000 cases and 1000 controls. The sample is analyzed using allele-based χ^2 association tests.

SOURCE CODE 5.11 Simulation of a Disease Model with G × G and G × E Interactions

```
import simuOpt
simuOpt.setOptions(gui=False, alleleType='binary')
import simuPOP as sim
import random, math

from ch6_genCaseCtrl import genCaseControlSample

def penetrance(alpha, beta1, beta2, beta3, gamma1, gamma2):
    def func(geno):
        e = random.randint(0, 1)
        g1 = geno[0] + geno[1]
        g2 = geno[2] + geno[3]
        logit = alpha + beta1*g1 + beta2*g2 + beta3*g1*g2 + gamma1*e*g1 + gamma2*e*g2
        return 1 / (1. + math.exp(-logit))
    return func

sample = genCaseControlSample(pop, 1000, 1000,
    sim.PyPenetrance(func=penetrance(-5, 0.20, 0.4, 0.4, 0.2, 0.4),
        loci=['rs4491689', 'rs6869003']))
sim.stat(sample, association=sim.ALL_AVAIL)
# get p-values
sample.dvars().Allele_ChiSq_p
```

We counted the number of alleles in cases and controls, created a 2×2 contingency table, and used a χ^2 test to assess the association between the disease status and alleles at each marker. The negative of the base 10 logarithm of the p-values at all markers were plotted in Figure 5.5. Although the χ^2 tests detected both genetic factors correctly, more sophisticated statistical methods or larger samples would be needed to detect the gene–gene and gene–environment interactions in this data set.

5.3.5 Example 3: Simulations of Slow and Rapid Selective Sweep

In the previous example, natural selection was not expected to have a strong impact on LD patterns around DPL because it changed frequencies of both DPAs slowly over a long period of time. In contrast, strong selection can bring a new beneficial mutation to high frequency or fixation in a population in a relatively short period of time. This phenomenon, also called selective sweep, has a profound impact on patterns of linked genetic variation through the hitchhiking effect. The signatures of selective sweep have been used for the development of statistical methods to identify chromosomal regions that have been under positive selection [32] and for the identification of DPL in genome-wide association analysis [33]. Although theo-

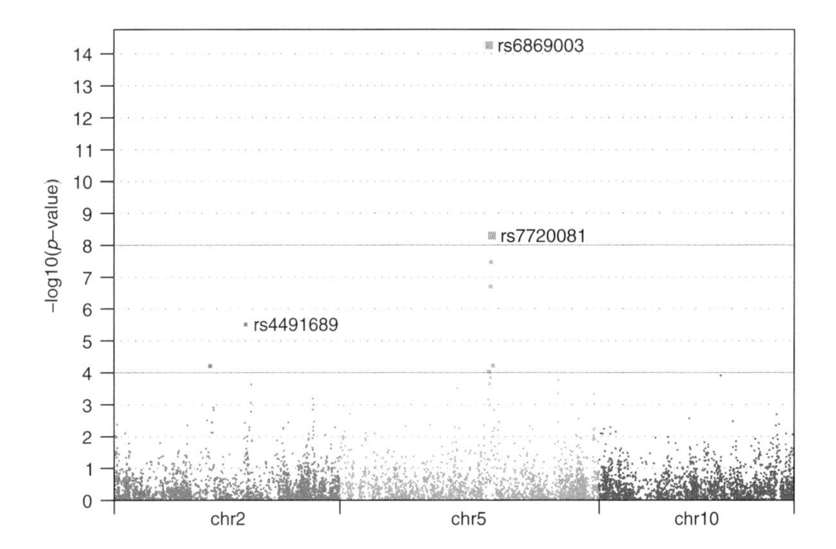

FIGURE 5.5 $-\text{Log}_{10}$ p-values of allele-based χ^2 tests at 6000 loci. Negative of the base 10 logarithm of p-values of allele-based χ^2 tests between 1000 cases and 1000 controls at 6000 markers (2000 each) on chromosomes 2, 5, and 10. Markers rs4491689 and rs6869003 are causal. Marker rs7720081 has low p-value because it is closely linked to marker rs6869003.

retical models of selective sweep have been simulated for methodological development using coalescent approaches [14, 32, 34], explicit simulation of selective sweep using a forward-time approach can be used to study the impact of different levels of natural selection on different regions of the human genome and to produce realistic samples to assess the power of statistical methods.

In order to closely examine the impact of selective sweep on existing LD patterns of a chromosomal region, we extracted 500 markers on a short region on chromosome 2 (chr2:234157787-234573235) from 170 independent individuals from the JPT+CHB population of phase 3 of the HapMap data set. We selected this region because it belongs to the ENr131 ENCODE region with a mean distance of 0.83 kb between markers. We selected marker rs2173746 with alleles C and T from one of the two haplotype blocks in this region and applied different levels of positive selection during the evolution of this population.

In our first simulation, we assumed that allele T at this marker was introduced more than 500 generations ago. This allele had a frequency of 5.88% in the initial HapMap population and reached a frequency of 99% after 500 generations due to positive natural selection with fitness 1,

1.02, and 1.03 for genotypes *CC*, *CT*, and *TT*, respectively. A forward-time trajectory simulation algorithm was used to control the frequency of allele *T* at the present generation. In our second simulation, we cleared allele *T* at this marker from the initial population and introduced it as a new mutant during the evolutionary process. Using a backward-time trajectory simulation process, an allele *T* was introduced at generation 268 and was brought to a frequency of 99% using a stronger force of natural selection with fitness 1, 1.05, and 1.11 for genotypes *CC*, *CT*, and *TT*, respectively. These simulations could be done through functions such as

```
simuGWAS(pop, DPL=['rs2173746'], curFreq=[0.99], fitness=[1, 1.07, 1.14])
```

with default values for other parameters.

We drew 1000 trios from the simulated populations using a rejection-sampling algorithm. More specifically, we repeatedly chose parents and produced offspring. We determined the affection status of each offspring using a logistic regression model

$$\text{logit}\left(\Pr\left(Y_1 = 1\right)\right) = -0.5 - g_i,$$

where g_i is the number of allele *T* at locus rs2173746. We kept only affected offspring and their parents in the sample until 1000 trios were collected.

■ EXAMPLE 5.12

This example implements a simple disease model with two parameters. In order to generate trio families with affected offspring, it defines a class `TioSampler` that evolves a population for one generation. Instead of using virtual subpopulations, this class uses an explicit function `_discardTrio` to discard offspring if he or she is unaffected or if his or her parents have already produced another affected offspring.

SOURCE CODE 5.12 Generation of Trio Samples from Simulated Population

```
import simuOpt
simuOpt.setOptions(gui=False, alleleType='binary')
import simuPOP as sim

import math
def penetrance(alpha, beta):
    def func(geno):
        g = geno[0] + geno[1]
        logit = alpha + beta*g
        return math.exp(logit) / (1. + math.exp(logit))
```

```
    return func

class TrioSampler:
    def __init__(self):
        # IDs of the parents of selected offspring
        self.parentalIDs = set()

    def _discardTrio(self, off):
        'Determine if the offspring can be kept.'
        if off.affected() and off.father_id not in self.parentalIDs and \
            off.mother_id not in self.parentalIDs:
            self.parentalIDs |= set([off.father_id, off.mother_id])
            return False
        # discard unaffected individual or individuals with duplicate parents
        return True

    def drawSample(self, pop, penet, nFamilies):
        self.pop = pop.clone()
        self.pop.addInfoFields(['ind_id', 'father_id', 'mother_id'])
        self.pop.setAncestralDepth(1)
        sim.tagID(self.pop, reset=True)
        self.pop.evolve(
            preOps = penet,
            matingScheme=sim.RandomMating(ops=[
                sim.MendelianGenoTransmitter(), # pass genotype
                sim.IdTagger(),        # assign new ID to offspring
                sim.PedigreeTagger(), # record the parent of each offspring
                penet,                 # determine offspring affection status
                sim.DiscardIf(cond=self._discardTrio)
                ], subPopSize=nFamilies),
            gen = 1
        )
        return self.pop

sample = TrioSampler().drawSample(pop,
    sim.PyPenetrance(penetrance(-0.5, -1.0), loci='rs2173746'), 1000)
```

We used LAMP [35] to analyze the data set. Figure 4 plots the simulated trajectories (Figure 5.6.b and c) as well as LD maps, drawn by HaploView [36], of this region before evolution for all 500 markers (Figure 4A) and after evolution with slow (Figure 5.6d) and rapid (Figure 5.6.e) sweeps for 100 markers around DPL. LD maps plot the pairwise D' measure of LD between all markers in a region, with high LD pairs marked in bright red. Comparing LD maps around DPL before (Figure 5.6a) and after evolution (Figure 5.6.d and e), it was clear that rapid selective sweep introduced blocks of monomorphic markers (gray areas around marker rs2173746 in Figure 5.6.e) when the haplotype with the mutant became prevalent when the mutant (allele T at marker rs2173746) was brought to a high frequency (99%). In contrast, the impact of slow selective sweep on LD patterns around DPL was barely discernible. As a matter of fact, the initial population had 50 haplotypes over a region of 100 markers around marker

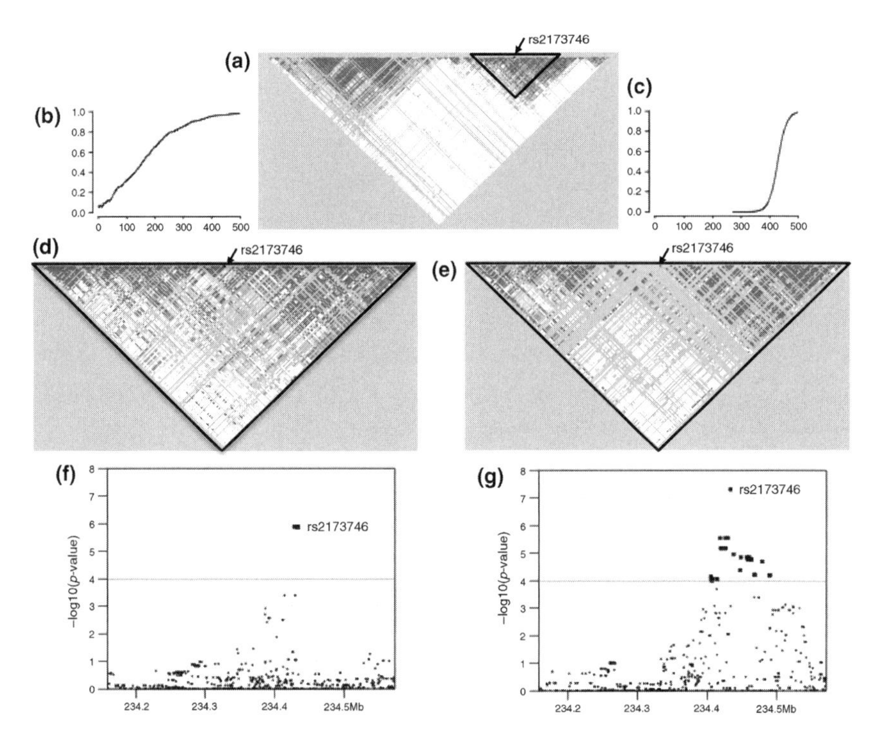

FIGURE 5.6 Impact of rapid and slow selective sweep on LD structure. An initial population of 170 independent individuals of JPT+CHB population of phase 3 of the HapMap data set was expanded to large populations and subjected to slow (b, d, and f) and rapid (c, e, g) selective sweeps at locus rs2173746. The trajectories of the frequencies of allele T at this marker in simulations after slow (b) and rapid (c) sweeps are plotted. LD maps of 500 markers on chromosome 2 of the initial population (a) and 100 markers around locus rs2173746 of expanded populations after the slow (d) and rapid (e) sweeps are plotted. From these expanded populations, 1000 cases and 1000 controls were drawn. The negative of the base 10 logarithm of p-values at 500 markers are plotted for slow (f) and rapid (g) sweeps.

rs2173746 (50 markers on each side), with a frequency of 16% for the most popular haplotype. After rapid selective sweep, only 19 haplotypes existed in this region, with a frequency of 97% for the most popular haplotype. In contrast, 195 haplotypes were present in the population resulting from slow selective sweep, with a frequency of 44% for the most popular haplotype.

The simulated populations could be used to test the performance of statistical tests designed to detect signals of positive selection along the human genome [32] and to detect DPL if DPA was under positive selection [33]. For example, when we applied family-based association tests to two samples of trio families drawn from the simulated populations, the signals from rapid selective sweep (Figure 5.6.g) appeared to be wider than those from slow selective sweep (Figure 5.6.f). This phenomenon could be explained by stronger LD between DPL and its surrounding loci for the simulation with rapid selective sweep, but a quantitative analysis using a large number of simulations would be needed to draw a definitive conclusion.

5.4 DISCUSSION

The genetic composition of a human population is the result of a long and complex evolutionary process. The demographic and genetic features of this process have profound implications in the mapping of susceptibility genes responsible for human genetic diseases. Although resampling-based methods capture the complexity of existing genome sequences with no control over the impact of additional genetic and demographic forces, and the coalescent methods simulate simple samples based solely on a few theoretical models, the forward-time simulation method described in this article retains key properties of human genomes by evolving a population from real human sequences while allowing the introduction of additional genetic forces such as natural selection. Because this method is not constrained by any theoretical limit, it can be used to simulate realistic samples for a variety of research topics for GWA studies.

In order to retain key features of real human genomes during evolution, this method expands the founder population rapidly. Because the size of the founder population is likely to be small, this evolutionary process currently suffers from a bottleneck effect during the initial stage of population expansion, resulting in a loss of rare haplotypes and reduced genotype diversity. During rapid population expansion, common haplotypes are maintained in the population with stable frequency, whereas new haplotypes are constantly introduced by mutation and recombination [30].

Consequently, common haplotypes in the initial population are preserved in the simulated population, but many rare haplotypes will be replaced by new haplotypes. Because mutation has a relatively small impact on common alleles, an increased mutation rate can be used to generate more new haplotypes in the simulated populations without markedly affecting other population properties such as allele frequency and LD patterns. This limitation will become less of a challenge as more human data become available (e.g., from 1000 Genomes Project [21]).

Because a larger initial population size would reduce the bottleneck effect and help preserve uncommon haplotypes, we combined all independent individuals from 10 populations of the phase 3 HapMap data set for examples 1, 2, and 4. The sudden population admixture caused long-range admixture LD in the combined initial population. Such admixture LD decayed gradually during evolution and did not lead to elevated long-range LD in the simulated population (Figure 5.2). Nevertheless, the availability of high-quality sequences of larger samples will allow to generate population-specific samples and further improve the quality of our simulated data sets.

We evaluated the quality of simulated data sets by comparing allele frequency and LD patterns between the simulated and the HapMap samples. However, we did not attempt to tweak our evolutionary process so that the simulated samples resembled the HapMap sample closely because we aimed to simulate larger populations with more genetic diversity than the founder population and because we wanted to use a realistic evolutionary process so that additional genetic or demographic features could be added. If we consider the method to randomly split and join pieces of chromosomes used by HapSample [37] as a special form of recombination, HapSample could be considered as one-generation forward-time simulation method with magnified recombination rates. Our simulation method would yield results similar to those of the resampling methods if we used an extraordinarily high scaling factor so that all sequences were essentially derived directly from haplotypes of the initial population.

We used a fine-scale genetic map to determine the recombination rate between adjacent markers [24]. This map generally has a higher recombination rate between markers from different haplotype blocks and a lower recombination rate between markers from the same haplotype blocks. Because recombinations happened mostly between existing haplotype blocks, this genetic map helped us retain the haplotype blocks and therefore LD structure of the founder population. However, due to the relatively short evolutionary time, the type of genetic map does not have a strong influence on the simulated population. For example, there is no discernible difference between LD patterns of simulated populations if we use a physical map with a recombination rate of 0.01 per Mbp instead of a genetic map to recombine parental chromosomes during evolution (results not shown).

Our simulation program allows the simulation of arbitrarily chosen selection models with multiple interacting genetic factors and at the mean time allele frequencies at the present generation. If inappropriate parameters are chosen, it is likely that the specified selection model would result in allele frequencies higher or lower than expected so that no valid trajectory of allele frequency could be simulated. If this is the case, our trajectory simulation function will print the average ending frequency for a forward-time simulation or length of trajectory for a backward-time simulation so that the simulation parameters can be adjusted accordingly. This is especially useful if a gene–gene interaction model is used so that marginal selection pressure can interact with allele frequency and drive the allele frequency of DPA in unpredictable directions.

Due to different requirements of different applications, the flexibility of this simulation method is difficult to harness using a traditional single-execution implementation. This is why we divided our simulation approach into three steps and implemented different preprocessing and postprocessing scripts for different applications. As we have demonstrated in a few examples, simuPOP can be used to produce different types of samples from a simulated population. This allows head-to-head comparison between not only statistical methods using the same simulated samples but also between statistical methods based on different study designs and sample types. The ability to simulate and study the entire population with disease is one of the greatest advantages of forward-time simulations.

REFERENCES

1. C. I. Amos, J. Krushkal, T. J. Thiel, A. Young, D. K. Zhu, E. Boerwinkle, and M. de Andrade, Comparison of model-free linkage mapping strategies for the study of a complex trait. *Genet Epidemiol*, 14(6):743–748, 1997.
2. P. C. Sham, S. Purcell, S. S. Cherny, and G. R. Abecasis, Powerful regression-based quantitative-trait linkage analysis of general pedigrees. *Am J Hum Genet*, 71(2):238–253, 2002.
3. J. Marchini, B. Howie, S. Myers, G. McVean, and P. Donnelly, A new multipoint method for genome-wide association studies by imputation of genotypes. *Nat Genet*, 39(7):906–913, 2007.
4. S. Wiltshire, A. P. Morris, and E. Zeggini, Examining the statistical properties of fine-scale mapping in large-scale association studies. *Genet Epidemiol*, 32(3):204–214, 2008.
5. C. C. A. Spencer, Z. Su, P. Donnelly, and J. Marchini, Designing genome-wide association studies: sample size, power, imputation, and the choice of genotyping chip. *PLoS Genet*, 5(5):e1000477, 2009.
6. H.-S. Chai, H. Sicotte, K. R. Bailey, S. T. Turner, Y. W. Asmann, and J.-P. A. Kocher, GLOSSI: a method to assess the association of genetic loci-sets with complex diseases. *BMC Bioinformatics*, 10:102, 2009.
7. H.-Y. Tan, J. H. Callicott, and D. R. Weinberger, Intermediate phenotypes in schizophrenia genetics redux: is it a no brainer? *Mol Psychiatry*, 13(3):233–238, 2008.
8. Z. Bochdanovits, M. Verhage, A. B. Smit, E. J. C. de Geus, D. Posthuma, D. I. Boomsma, B. W. J. H. Penninx, W. J. Hoogendijk, and P. Heutink,

Joint reanalysis of 29 correlated SNPS supports the role of PCLO/Piccolo as a causal risk factor for major depressive disorder. *Mol Psychiatry*, 14(7): 650–652, 2009.

9. Y. Liu, G. Athanasiadis, and M. E. Weale, A survey of genetic simulation software for population and epidemiological studies. *Hum Genomics*, 3(1): 79–86, 2008.
10. G. Coop and R. C. Griffiths, Ancestral inference on gene trees under selection. *Theor Popul Biol*, 66(3):219–232, 2004.
11. Y. Wang and B. Rannala, *In silico* analysis of disease-association mapping strategies using the coalescent process and incorporating ascertainment and selection. *Am J Hum Genet*, 76(6):1066–1073, 2005.
12. M. Slatkin, Simulating genealogies of selected alleles in a population of variable size. *Genet Res*, 78(1):49–57, 2001.
13. T. Maruyama, The age of a rare mutant gene in a large population. *Am J Hum Genet*, 26(6):669–673, 1974.
14. C. C. A. Spencer and G. Coop, SelSim: a program to simulate population genetic data with natural selection and recombination. *Bioinformatics*, 20(18):3673–3675, 2004.
15. T. Mailund, M. H. Schierup, C. N. S. Pedersen, P. J. M. Mechlenborg, J. N. Madsen, and L. Schauser, CoaSim: a flexible environment for simulating genetic data under coalescent models. *BMC Bioinformatics*, 6:252, 2005.
16. B. Peng, C. I. Amos, and M. Kimmel, Forward-time simulations of human populations with complex diseases. *PLoS Genet*, 3(3):e47, 2007.
17. T. Mehta, M. Tanik, and D. B. Allison, Towards sound epistemological foundations of statistical methods for high-dimensional biology. *Nat Genet*, 36(9):943–947, 2004.
18. D. Reich and N. Patterson, Will admixture mapping work to find disease genes? *Philos Trans R Soc Lond B Biol Sci*, 360(1460):1605–1607, 2005.
19. International HapMap Consortium, A haplotype map of the human genome. *Nature*, 437(7063):1299–1320, 2005.
20. M. I. McCarthy, G. R. Abecasis, L. R. Cardon, D. B. Goldstein, J. Little, J. P. A. Ioannidis, and J. N. Hirschhorn, Genome-wide association studies for complex traits: consensus, uncertainty and challenges. *Nat Rev Genet*, 9(5):356–369, 2008.
21. J. Wise, Consortium hopes to sequence genome of 1000 volunteers. *BMJ*, 336(7638):237, 2008.
22. J. D. Wall and M. Przeworski, When did the human population size start increasing? *Genetics*, 155(4):1865–1874, 2000.
23. W. J. Ewens, "Mathematical Population Genetics", 2004, Springer.
24. S. Myers, L. Bottolo, C. Freeman, G. McVean, and P. Donnelly, A fine-scale map of recombination rates and hotspots across the human genome. *Science*, 310(5746):321–324, 2005.

25. M. Kimura and G. H. Weiss, The stepping stone model of population structure and the decrease of genetic correlation with distance. *Genetics*, 49(4):561–576, 1964.
26. C. J. Hoggart, M. Chadeau-Hyam, T. G. Clark, R. Lampariello, J. C. Whittaker, M. De Iorio, and D. J. Balding, Sequence-level population simulations over large genomic regions. *Genetics*, 177(3):1725–1731, 2007.
27. D. Ford, D. F. Easton, M. Stratton, S. Narod, D. Goldgar, P. Devilee, D. T. Bishop, B. Weber, G. Lenoir, J. Chang-Claude, H. Sobol, M. D. Teare, J. Struewing, A. Arason, S. Scherneck, J. Peto, T. R. Rebbeck, P. Tonin, S. Neuhausen, R. Barkardottir, J. Eyfjord, H. Lynch, B. A. Ponder, S. A. Gayther, and M. Zelada-Hedman, Genetic heterogeneity and penetrance analysis of the BRCA1 and BRCA2 genes in breast cancer families: the breast cancer linkage consortium. *Am J Hum Genet*, 62(3):676–689, 1998.
28. M. Slatkin, Linkage disequilibrium in growing and stable populations. *Genetics*, 137(1):331–336, 1994.
29. G. A. T. McVean, A genealogical interpretation of linkage disequilibrium. *Genetics*, 162(2):987–991, 2002.
30. B. Peng, R. K. Yu, K. L. Dehoff, and C. I. Amos, Normalizing a large number of quantitative traits using empirical normal quantile transformation. *BMC Proc*, 1 (Suppl 1):S156, 2007.
31. D. Li and D. V. Conti, Detecting gene–environment interactions using a combined case-only and case-control approach. *Am J Epidemiol*, 169(4):497–504, 2009.
32. B. F. Voight, S. Kudaravalli, X. Wen, and J. K. Pritchard, A map of recent positive selection in the human genome. *PLoS Biol*, 4(3):e72, 2006.
33. G. Ayodo, A. L. Price, A. Keinan, A. Ajwang, M. F. Otieno, A. S. S. Orago, N. Patterson, and D. Reich, Combining evidence of natural selection with association analysis increases power to detect malaria-resistance variants. *Am J Hum Genet*, 81(2):234–242, 2007.
34. G. McVean, The structure of linkage disequilibrium around a selective sweep. *Genetics*, 175(3):1395–1406, 2007.
35. M. Li, M. Boehnke, and G. R. Abecasis, Joint modeling of linkage and association: identifying SNPS responsible for a linkage signal. *Am J Hum Genet*, 76(6):934–949, 2005.
36. J. C. Barrett, B. Fry, J. Maller, and M. J. Daly, Haploview: analysis and visualization of ld and haplotype maps. *Bioinformatics*, 21(2):263–265, 2005.
37. F. A. Wright, H. Huang, X. Guan, K. Gamiel, C. Jeffries, W. T. Barry, F. P.-M. de Villena, P. F. Sullivan, K. C. Wilhelmsen, and F. Zou, Simulating association studies: a data-based resampling method for candidate regions or whole genome scans. *Bioinformatics*, 23(19):2581–2588, 2007.

CHAPTER 6

NONRANDOM MATING AND ITS APPLICATIONS

In the previous chapters, we have mostly used a random mating scheme to model the evolution of human populations. The essential feature of this mating scheme is that if we do not consider natural selection, the probability that two individuals in a population will mate is the same for all possible pairs of individuals. Although not all parents will pass their genotypes to the offspring generation (see Section 2.7 for an analysis of the genealogy of forward-time simulations), genotypes of the parental population are mixed efficiently among members of the next generation. Consequently, an evolutionary process using this mating scheme has a relatively large effective population size.

However, nonrandom mating can be important in studies of some populations of plants and animals, and even humans. In contrast to a random mating scheme that can be defined rigorously, deviation from random mating can take many different forms. For example, many plants reproduce by varying degrees of self-fertilization. Different levels of inbreeding exists in animal populations, animal populations have many different mating patterns that cannot be modeled by a random mating scheme. If we limit ourselves to sexual mating in diploid populations, nonrandom mating occurs when the probability that two individuals in a population will mate is

Forward-time Population Genetics Simulations: Methods, Implementation, and Applications,
Bo Peng, Marek Kimmel, and Christopher I. Amos.

not the same for all possible pairs of individuals. This is certainly the case for human populations because mating decisions of humans are usually influenced traits such as appearance, personality, cultural values, and social status. As we will demonstrate in this chapter, nonrandom mating has strong influence on the evolution of genes that are related to these traits and should be considered when we study the genetic background of these traits. Note that the mating patterns we consider here are trait specific because traits that are unrelated to mating (e.g., hypertension) can flow roughly freely during evolution. Even if the population mates nonrandomly, genotypes that are associated with these traits can be considered to be under random mating.

This chapter demonstrates how to simulate arbitrary nonrandom mating schemes in human and other populations using simuPOP. In addition to usual nonrandom mating schemes such as assortative mating, we also demonstrate how to use simuPOP to simulate age-structured populations using overlapping generations. Examples in this chapter use some advanced features of simuPOP, so refer to Section 6.10 if you are interested in the implementation of these examples.

6.1 ASSORTATIVE MATING

Assortative mating refers to mate selection on the basis of phenotypic or genotypic characteristics. The most common form of assortative mating among humans is one in which individuals mate with others who are like themselves phenotypically for selected traits. This is referred to as *positive assortative mating*. For example, the marital correlation (correlation of a trait among spouses) of body mass index (BMI) is 0.13 [1], and is as high as 0.26 for systolic blood pressure [2]. Although such correlations can be caused by other effects such as increased spousal similarity with age or cohabitation, evidence has shown that some correlations are truly a function of assortative mating.

Assortative mating can have a strong impact on the evolution of genes that are related to the traits. Using an extreme example, if we assume that a trait with three values is caused by the number of disease allele *a* at a single locus and people only mate with spouses with an identical trait, only three possible mating patterns will exist in the population, namely, *AA* with *AA*, *Aa* with *Aa*, and *aa* with *aa*. This will cause an increase of homozygous genotypes (*AA* and *aa*) and a decrease of heterozygous ones (*Aa*) in a population, which is a deviation of Hardy–Weinberg equilibrium at this locus.

Assortative mating can also cause spurious associations in genetic association studies. For example, Redden and Allison [3] used forward-time simulations to simulate populations that have gone through positive assortative mating with respect to adiposity, beauty, and intelligence and studied the impact of such nonrandom mating on the type I crror of several statistical gene mapping methods. This section implements this simulation study to demonstrate how to use a homogeneous mating scheme to simulate assortative mating.

6.1.1 Genetic Architecture of Traits

Quantitative traits refer to phenotypes (properties) that vary in degree in a population. They can be binary (e.g., onset of a disease), categorical (e.g., race), or numeric (e.g., blood pressure). A quantitative trait model determines one or more traits of individuals according to his or her genotype, sex, or other properties. Quantitative traits of individuals are usually assigned during evolution to model the evolution of traits or after evolution for, for example, sampling purposes. Example 6.1 demonstrates how to apply a penetrance model to a population in simuPOP.

Redden and Allison [3] simulated 40 independent unlinked diallelic loci for all simulated individuals. An allele frequency of 0.5 was used for all allele, which led to genotype frequencies 0.25, 0.50, and 0.25 for genotypes *AA*, *Aa*, and *aa*. These 40 loci are divided into 4 groups, among which 10 of the loci (labeled a1–a10) were used to simulate the phenotype adiposity (*A*), 10 loci (labeled b1–b10) were used to simulate the phenotype beauty (*B*), and 10 loci (labeled i1–i10) were used to simulate the phenotype intelligence (*I*). The final 10 loci (labeled u1–u10) were designated as uninformative "random markers" and were not utilized in the construction of any phenotype.

The phenotypic contribution of genotypes in these groups to the corresponding phenotype is simply the number of *a* alleles at these loci. A normal random variable with mean of 0 and variance 2.5 was added to that sum to create the phenotypic distributions.

■ EXAMPLE 6.1

simuPOP stores trait values as floating point numbers in information fields. A standard method to implement a quantitative trait model in simuPOP is to implement it in a callback function and pass the function to a `PyQuanTrait` operator. This callback function accepts one or more parameters of `geno` (genotype at specified loci), `gen` (generation number), `ind` (individual), or names of information fields, and return one or more trait values. For example, a callback function

```
def traits(geno):
    return sum(geno[:10]) + random.normalvariate(0, 2.5), \
        sum(geno[10:20]) + random.normalvariate(0, 2.5), \
        sum(geno[20:30]) + random.normalvariate(0, 2.5),
```

defines a quantitative trait model for Redden and Allison [3]. When this function is used in a `PyQuanTrait` operator, for example,

```
PyQuanTrait(func=traits, loci=range(30), infoFields=['A', 'B', 'I'])
```

the operator will collect genotype at specified loci for each individual, pass them to this function, and assign the return values to specified information fields *A*, *B*, and *I*.

If additional information is needed in a quantitative trait, it can be passed to the callback function as parameters. For example, function

```
def qtrait(geno, ind):
    'A sex-dependent quantitative trait model'
    if ind.sex() == MALE:
        return random.normalvariate(sum(geno), 1)
    else:
        return random.normalvariate(sum(geno), 0.5)
```

defines a quantitative trait model that returns a random trait that follows a Gaussian distribution with mean 0, 1, and 2 for genotypes (0,0), (0,1), and (1,1), respectively, and a variance 1 for male individuals and 0.5 for female individuals. When this function is passed to operator

```
PyQuanTrait(func=qtrait, loci=3, infoFields='trait')
```

this operator will go through all individuals, pass genotype at locus 3 and the individual object itself to function `qtrait`, and assign its return value to information field `trait`.

This example demonstrates a quantitative trait model that depends on an information field `age`. This model is defined in function `qtrait` that accepts parameters `geno` and `age` and returns a tuple of two elements. The return values are used to assign trait values to information fields `trait1` and `trait2`. Although it is more realistic to apply this quantitative trait model to an age-structured population, this example assigns individual fitness randomly for the sake of simplicity.

SOURCE CODE 6.1 A Quantitative Trait Model

```
>>> import simuPOP as sim
>>> import random
>>> pop = sim.Population(size=5000, loci=2, infoFields=['qtrait1', 'qtrait2', 'age'])
>>> pop.setVirtualSplitter(sim.InfoSplitter(field='age', cutoff=[40]))
>>> def qtrait(geno, age):
...     'Return two traits that depends on genotype and age'
...     return random.normalvariate(age * sum(geno), 10), random.randint(0, 10*sum(geno))
...
>>> pop.evolve(
...     initOps=[
...         sim.InitSex(),
...         sim.InitGenotype(freq=[0.2, 0.8]),
...     ],
...     matingScheme=sim.RandomMating(),
...     postOps=[
...         # use random age for simplicity
...         sim.InitInfo(lambda:random.randint(20, 75), infoFields='age'),
...         sim.PyQuanTrait(loci=(0,1), func=qtrait, infoFields=['qtrait1', 'qtrait2']),
...         sim.Stat(meanOfInfo=['qtrait1'], subPops=[(0, sim.ALL_AVAIL)],
...             vars='meanOfInfo_sp'),
...         sim.PyEval(r"'Mean of trait1: %.3f (age < 40), %.3f (age >=40)\n' % "
...             "(subPop[(0,0)]['meanOfInfo']['qtrait1'], subPop[(0,1)]['meanOfInfo']['qtrait1'])"),
...     ],
...     gen = 5
... )
Mean of trait1: 93.566 (age < 40), 182.453 (age >=40)
Mean of trait1: 94.545 (age < 40), 182.620 (age >=40)
Mean of trait1: 95.227 (age < 40), 183.338 (age >=40)
Mean of trait1: 95.808 (age < 40), 183.537 (age >=40)
Mean of trait1: 94.684 (age < 40), 185.260 (age >=40)
5
>>>
```

6.1.2 Mating Model

Redden and Allison [3] used a simple model of mate selection in which people mated assortatively on the basis of an emergent trait D (D for desirability). In this model, $D = B + I - A + \varepsilon$, where ε is a normally distributed random variable with mean zero and variance σ_{ε}^2. The value of σ_{ε} was used to control the level of assortative mating.

To simulate assortative mating, Redden and Allison [3] sorted males and females independently by D and then paired upon D such that the highest ranked male mated with the highest ranked female and so on for each successive rank. Although the pairings of parents are performed sequentially, there are varying levels of randomness for the pairing of parents with similar genotypes because traits A, B, I, and D all have a random component. In order to keep constant population size and have exact numbers of males and females in the population, all mating events produce a male and a female offspring. In summary, we need a mating scheme that involves the following:

- Match males and females by their desirability values. That is to say, a male with the highest desirability value in male individuals mates with a female with the highest desirability value in female individuals, a male with the second highest desirability value mates with a female with the second highest desirability value, and so on.
- Produce equal numbers of male and female individuals in the offspring population.

We simulate this mating scheme by sorting individuals by their desirability (D) values and select and mate parents sequentially. Because parents can only mate once, we produce two offspring, one male and one female, per mating event to keep a constant population size and keep equal numbers of males and females. This is explained in the following example.

■ EXAMPLE 6.2

In order to line up males and females side by side, it is necessary to sort individuals according to trait value D. This can be easily achieved using function `Population.sortIndividuals`. However, because this is a member function that sorts individuals of a particular population, it cannot be used as an operator during evolution (because operators are objects that are applied to populations repeatedly). The solution is to define a function

```
def sortByD(pop):
    pop.sortIndividuals('D')
    return True
```

and use it in a Python operator

```
PyOperator(func=sortByD)
```

This operator needs to be applied immediately before mating.

Because there is no predefined mating scheme that selects parents sequentially, we will need to define our own mating scheme. This is not particularly difficult, but implementing this mating scheme requires some understanding of the homogeneous mating scheme of simuPOP. Homogeneous mating schemes in simuPOP refer to mating schemes that use a single method to choose parents and produce offspring. It is composed of a *parent chooser* that is responsible for choosing one or two parents from the parental generation and an *offspring generator* that generates one or more offspring from the chosen parents. For example, the most commonly used mating scheme, namely, the diploid sexual RandomMating mating scheme, is defined as

```
def RandomMating(numOffspring=1, sexMode=RANDOM_SEX, ops=MendelianGenoTransmitter(),
    subPopSize=[], subPops=ALL_AVAIL, weight=0, selectionField='fitness'):
    return HomoMating(chooser=RandomParentsChooser(True, selectionField),
        generator=OffspringGenerator(ops, numOffspring, sexMode),
        subPopSize=subPopSize, subPops=subPops, weight=weight)
```

That is to say, this mating scheme is a homogeneous mating scheme (HomoMating) that uses a RandomParentsChooser to choose two parents with replacement (the first parameter) from their respective sex group and an OffspringGenerator to produce offspring. A genotype transmitter MendelianGenoTransmitter is used by default to transmit genotypes from parents to offspring. Parameters numOffspring and sexMode are passed to this offspring generator to control the number of offspring per mating event and the sex of offspring. The constructor of a homogeneous mating scheme also accepts parameters subPopSize, subPops, and weight. Parameter subPopSize is used to control the size of the offspring subpopulation. The latter two parameters are significant only when the mating scheme is used in a heterogeneous mating scheme, which will be discussed later.

simuPOP provides a number of genotype transmitters, parent choosers, and offspring generators and uses them to define a number of commonly used homogeneous mating schemes (Table 6.1). If a special homogeneous mating scheme is needed, it can usually be defined using existing parent choosers and offspring generators. For example, a SelfMating mating scheme uses a RandomParentChooser to choose a parent randomly from a population and a selfingOffspringGenerator to transmit his or her genotype to his or her offspring. If a theoretical model requires that all parents transmit their genotypes to an offspring population, you can use a SequentialParentChooser to choose parents sequentially and self-fertilize parents one by one.

An example of a customized homogeneous mating scheme is provided in this example. This example uses a mating scheme where a sequential parent chooser is used to select parents one by one. As you can see from the value of parent_idx recorded by a ParentsTagger, the first offspring inherits his genotype from the first parent, the second offspring from the second parent, and so on. Because of the use of a selfing genotype

TABLE 6.1 Mating Schemes-Related Classes

Object	Usage
Genotype transmitters	
`CloneGenoTransmitter`	Copy genotype, sex, and all information fields of a parent to offspring
`MendelianGenoTransmitter`	Select one of two parental chromosomes of two parents randomly and pass them to the offspring
`SelfingGenoTransmitter`	Select one of two parental chromosomes randomly of a parent twice and pass them to the offspring
`HaplodiploidGenoTransmitter`	Copy the first set of chromosomes of a male parent and a random set of chromosomes of a female parent to offspring
`MitochondrialGenoTransmitter`	Copy one or more customized chromosomes of the female parent to offspring
`Recombinator`	Recombine parental chromosomes and copy one of the recombinants to offspring
Parent choosers	
`SequentialParentChooser`	Choose a parent sequentially
`SequentialParentChooser`	Choose a male and a female parent sequentially from their respective sex group
`RandomParentChooser`	Choose a parent randomly, regardless of sex, with or without replacement
`RandomParentsChooser`	Choose a male and a female parent randomly, with or without replacement
`PolyParentsChooser`	Similar to a random parents chooser, but one of the parents will mate with several spouses before he or she is replaced

(*Continued*)

TABLE 6.1 Mating Schemes-Related Classes (*Continued*)

Object	Usage
`CombinedParentsChooser`	Return a pair of parents from two parent choosers
`PyParentsChooser`	Return a parent or two parents by calling a user-defined Python generating function
Offspring generators	
`OffspringGenerator`	Produce offspring from one or two parents using the provided during-mating operators
`ControlledOffspringGenerator`	Selectively accepting offspring to control the frequencies of alleles at one or more loci
Predefined mating schemes	
`CloneMating`	Select parents sequentially and clone them to the offspring population
`RandomSelection`	An asexual haploid random mating scheme
`RandomMating`	A diploid random mating scheme
`MonogamousMating`	A random mating scheme in which parents are chosen only once
`HaplodiploidMating`	A mating scheme in haplodiploid populations, using a genotype transmitter designed for such a population
`SelfMating`	A random selection mating scheme that uses a selfing genotype transmitter
`ControlledRandomMating`	A random mating scheme using a controlled offspring generator

transmitter, parental chromosomes are chosen twice (randomly) to form the two homologous copies of offspring chromosomes.

With the appropriate parent chooser and offspring generator to use, it is then easy to translate the required mating scheme into the simuPOP language:

```
HomoMating(
    chooser=SequentialParentsChooser(),
    generator=OffspringGenerator(ops=MendelianGenoTransmitter(),
        numOffspring=2, sexMode=(NUM_OF_MALES, 1))
)
```

Here we use a `sexMode` that specifies one and only one male individual in a family, so there will be a male and a female when `numOffspring=2`.

SOURCE CODE 6.2 A Sequential Selfing Mating Scheme

```
>>> import simuPOP as sim
>>> pop = sim.Population(100, loci=[5]*3, infoFields='parent_idx')
>>> pop.evolve(
...     initOps=sim.InitGenotype(freq=[0.2]*5),
...     preOps=sim.Dumper(structure=False, max=5),
...     matingScheme=sim.HomoMating(
...         sim.SequentialParentChooser(),
...         sim.OffspringGenerator(ops=[
...             sim.SelfingGenoTransmitter(),
...             sim.ParentsTagger(infoFields='parent_idx'),
...         ])
...     ),
...     postOps=sim.Dumper(structure=False, max=5),
...     gen = 1
... )
SubPopulation 0 (), 100 Individuals:
   0: MU 44100 01422 24423 | 43130 34400 10114 |  0
   1: MU 33444 24430 34342 | 11320 34413 33201 |  0
   2: MU 03434 40424 24240 | 34430 41214 30212 |  0
   3: MU 13232 23304 20043 | 14130 02231 14240 |  0
   4: MU 11112 30400 33342 | 34434 42211 33120 |  0

SubPopulation 0 (), 100 Individuals:
   0: MU 44100 34400 24423 | 44100 34400 10114 |  0
   1: MU 33444 24430 33201 | 33444 24430 33201 |  1
   2: FU 34430 41214 24240 | 34430 41214 24240 |  2
   3: FU 13232 02231 14240 | 14130 02231 20043 |  3
   4: MU 34434 30400 33342 | 11112 42211 33342 |  4

1
```

6.1.3 Simulation of Assortative Mating

Example 6.3 demonstrates how to simulate the assortative mating scheme described in Redden and Allison [3]. Although Redden and Allison [3] used different statistical methods to test and correct for spurious associations caused by assortative mating, we simply calculate the Pearson correlation between genotypes at the *A* loci (loci 1 through 10) with traits *A* and *B* and genotypes at the *U* loci (loci 31 through 40) with trait *A*.

As we can see from the correlation values, trait *A* is strongly correlated with genotypes at the *A* loci, but not correlated with genotypes at the *U* loci. This is expected because the *A* loci contribute directly to trait *A*, and the *U* loci are not related to any trait. What is interesting is the apparent correlation between trait *B* and genotypes at the *A* loci. Although these loci are not genetically related to trait *B*, they contribute to trait *D* by which assortative mating occurs. Such correlations are therefore an artifact of positive assortative mating.

■ EXAMPLE 6.3

This example starts with a definition of a quantitative trait function. Instead of deriving trait *D* from trait values *A*, *B*, and *I*, this function handles all traits together in function `traits`. This function is used in a `PyQuanTrait` operator to calculate traits *A*, *B*, *I*, and *D* for each individual at the beginning of each generation.

The simulation starts with a founding population of 100,000 individuals. Because our mating scheme requires equal numbers of male and female individuals, we use a `InitSex` operator with parameter `maleProp=0.5` to specify the proportion of male individuals in the population. This results in a population of 50,000 men and 50,000 women. All loci in this population are initialized with allele frequency 0.5.

Individuals are sorted by their desirability values before mating starts at each generation. Because the function to sort individual is very short, this example uses a lambda function to define the function in place. Note that a callback function for a `PyOperator` must return `True` or `False`, so this lambda uses `is None` after `pop.sortIndividuals('D')` to return `True` because `sortIndividuals()` does not return a value.

After individuals are sorted, a positive assortative mating scheme is used to evolve the population for 10 generations. After 10 generations, the simulated population are analyzed for spurious associations. Because there is no built-in function to calculate correlation between two vectors, this example uses a `rpy` module to pass vectors to a statistical package R and calculate correlation using a *R* function `cor`.

SOURCE CODE 6.3 An Example of Assortative Mating

```
>>> import simuPOP as sim
>>> from random import normalvariate
>>> sigma = 1
>>> def traits(geno):
...     'genotypes are arranged as a1a2b1b2c1c2... where a,b,c are specified loci'
...     A = sum(geno[:20]) + normalvariate(0, 2.5)
```

```
...     B = sum(geno[20:40]) + normalvariate(0, 2.5)
...     I = sum(geno[40:60]) + normalvariate(0, 2.5)
...     D = B + I - A + normalvariate(0, sigma**2)
...     return A, B, I, D
...
>>> pop = sim.Population(100000, loci=[1]*40, infoFields=['A', 'B', 'I', 'D'])
>>> pop.evolve(
...     initOps=[
...         sim.InitSex(maleProp=0.5),
...         sim.InitGenotype(freq=[0.5, 0.5]),
...     ],
...     preOps=[
...         sim.PyQuanTrait(func=traits, loci=sim.ALL_AVAIL,
...             infoFields=['A', 'B', 'I', 'D']),
...         sim.PyOperator(func=lambda pop: pop.sortIndividuals('D') is None),
...     ],
...     matingScheme=sim.HomoMating(
...         chooser=sim.SequentialParentsChooser(),
...         generator=sim.OffspringGenerator(
...             ops=sim.MendelianGenoTransmitter(),
...             numOffspring=2, sexMode=(sim.NUM_OF_MALES, 1))
...     ),
...     finalOps=sim.PyQuanTrait(func=traits, loci=sim.ALL_AVAIL,
...             infoFields=['A', 'B', 'I', 'D']),
...     gen=10
... )
10
>>>
>>> from rpy import r
>>> def genoTraitCorrelation(loc, trait):
...     'Calculate correlation between trait and genotype at a locus'
...     geno = [ind.allele(loc,0) + ind.allele(loc,1) for ind in pop.individuals()]
...     qtrait = pop.indInfo(trait)
...     return r.cor(geno, qtrait)
...
>>> # correlation between genotype at A loci with trait A
>>> AA = [genoTraitCorrelation(loc, 'A') for loc in range(10)]
>>> print(', '.join(['%.3f' % abs(x) for x in AA]))
0.267, 0.270, 0.275, 0.272, 0.273, 0.272, 0.273, 0.273, 0.271, 0.273
>>> # correlation between genotype at A loci with trait B (spurious)
>>> AB = [genoTraitCorrelation(loc, 'B') for loc in range(10)]
>>> print(', '.join(['%.3f' % abs(x) for x in AB]))
0.082, 0.086, 0.087, 0.080, 0.084, 0.086, 0.088, 0.081, 0.091, 0.086
>>> # correlation between genotype at unrelated loci with trait A
>>> UA = [genoTraitCorrelation(loc, 'A') for loc in range(30, 40)]
>>> print(', '.join(['%.3f' % abs(x) for x in UA]))
0.002, 0.003, 0.000, 0.001, 0.004, 0.001, 0.005, 0.006, 0.000, 0.001
```

6.2 MORE COMPLEX NONRANDOM MATING SCHEMES

6.2.1 Customized Parent Choosing Scheme

A parent choosing scheme can be quite complicated in reality. For example, long-finned pilot whales swim in large social groups known as pods, but male whales neither disperse from nor mate within their natal pods. Instead, they will temporarily leave their pod and mate with females in another pod [4]. Example 6.4 implements the mating behavior of pilot whales by explicitly choosing parents in a Python function. Although this method is not particularly efficient, it allows the implementation of arbitrarily complex mating schemes. Methods to improve the efficiency of this method is discussed in the simuPOP users' guide.

■ EXAMPLE 6.4

A hybrid parent chooser `PyParentsChooser` accepts a user-defined Python generator function. This generator function takes a population and a subpopulation index as parameters `pop` and `subPop`. When this parent chooser is applied to a subpopulation, it will call this generator function and ask repeatedly for either a parent or a pair of parents. References to both individual objects or indices relative to a subpopulation are acceptable. Because a generator function usually do not know the population size of the offspring population and thus the numbers of parents to return, a `while True` loop is usually used to yield parents indefinitely. Because this function will not be called after the offspring population has been filled, this infinite loop will not be executed indefinitely.

This example creates a population of 5000 whales. Because mating can happen only within a subpopulation and a male parent will not leave his pod after a mating event, this example does not use subpopulations to represent different pods. Instead, it uses an information field `pod` to identify the pod each individual belongs to.

The way male whales mate with female whales in another pod is implemented in function `podParentsChooser`. In this function, male and female whales are separated into lists of males and females. Whenever a pair of parents is needed, a male whale is chosen randomly and female whales are chosen repeatedly until one in a different pod with the chosen male is identified.

This example uses a number of advanced features of simuPOP to simulate and observe the evolution of disease alleles at two loci, one on an autosomal chromosome and one on a mitochondrial chromosome. The type of each chromosome is specified by parameter `chromTypes`. The mitochondrial chromosome is denoted as a `CUSTOMIZED` type. Such chromosomes are ignored by regular genotype transmitters, but a `MitochondrialGenoTransmitter` will choose customized chromosomes of a female parent randomly and pass them to an offspring. This is why two genotype transmitters, one for each chromosome, are used in the `ops` parameter of an `OffspringGenerator`. Because newborns stay with their natal pod, we use an `InheritTagger` to assign each offspring with a pod index copied from their mother.

We assume that only the first pod has disease alleles and use an operator `InitGenotype` to initialize two loci with allele frequency 0.2 to whales belonging to the first virtual subpopulation (VSP). Because male whales visit other pods and leave their genotype in other pods and female whales stay at their own pod, the disease allele at the first locus spreads to other pods quickly. Because mitochondrial chromosomes are transmitted maternally, the disease allele at the second locus stays in the first pod during evolution.

SOURCE CODE 6.4 Simulation of Mating Behaviors of Pilot Whales

```
>>> import simuPOP as sim
>>> from random import randint
>>>
>>> def podParentsChooser(pop, subPop):
...     '''Choose parents of parents from different pods'''
...     males = [x for x in pop.individuals(subPop) if x.sex() == sim.MALE]
...     females = [x for x in pop.individuals(subPop) if x.sex() == sim.FEMALE]
...     while True:
...         # randomly choose a male
...         male = males[random.randint(0, len(males)-1)]
...         pod = male.pod
...         # randomly choose a female from different pod
...         while True:
...             female = females[randint(0, len(females)-1)]
...             if female.pod != pod:
...                 break
...         yield (male, female)
...
>>> pop = sim.Population(5000, loci=[1,1], infoFields=['pod'],
...     chromTypes=[sim.AUTOSOME, sim.CUSTOMIZED])
>>> pop.setVirtualSplitter(sim.InfoSplitter('pod', values=range(5)))
>>> pop.evolve(
...     initOps = [
...         sim.InitSex(),
...         # assign individuals to a random pod
...         sim.InitInfo(lambda : randint(0, 4), infoFields='pod'),
...         # only the first pod has the disease alleles
...         sim.InitGenotype(freq=[0.8, 0.2], subPops=[(0,0)]),
...     ],
...     matingScheme = sim.HomoMating(
...         sim.PyParentsChooser(podParentsChooser),
...         sim.OffspringGenerator(numOffspring=1, ops=[
...             sim.MendelianGenoTransmitter(),
...             sim.MitochondrialGenoTransmitter(),
...             # offspring stays with their natal pod
...             sim.InheritTagger(mode=sim.MATERNAL, infoFields='pod')])),
...     postOps = [
...         # calulate allele frequency at each pod
```

```
...         sim.Stat(alleleFreq=(0,1), vars='alleleFreq_sp',
...             subPops=[(0, sim.ALL_AVAIL)]),
...         sim.PyEval(r"'Loc0: %s Loc1: %s\n' % ("
...     "', '.join(['%.3f' % subPop[(0,x)]['alleleFreq'][0][1] for x in range(5)]),"
...     "', '.join(['%.3f' % subPop[(0,x)]['alleleFreq'][1][1] for x in range(5)]))"),
...     ],
...     gen = 10
... )
Loc0: 0.086, 0.021, 0.026, 0.025, 0.022 Loc1: 0.108, 0.000, 0.000, 0.000, 0.000
Loc0: 0.055, 0.028, 0.027, 0.033, 0.032 Loc1: 0.103, 0.000, 0.000, 0.000, 0.000
Loc0: 0.039, 0.030, 0.033, 0.023, 0.031 Loc1: 0.094, 0.000, 0.000, 0.000, 0.000
Loc0: 0.042, 0.034, 0.030, 0.024, 0.025 Loc1: 0.092, 0.000, 0.000, 0.000, 0.000
Loc0: 0.036, 0.036, 0.021, 0.033, 0.026 Loc1: 0.102, 0.000, 0.000, 0.000, 0.000
Loc0: 0.023, 0.030, 0.032, 0.029, 0.026 Loc1: 0.085, 0.000, 0.000, 0.000, 0.000
Loc0: 0.032, 0.034, 0.030, 0.028, 0.021 Loc1: 0.095, 0.000, 0.000, 0.000, 0.000
Loc0: 0.036, 0.026, 0.027, 0.036, 0.026 Loc1: 0.118, 0.000, 0.000, 0.000, 0.000
Loc0: 0.036, 0.027, 0.023, 0.037, 0.024 Loc1: 0.126, 0.000, 0.000, 0.000, 0.000
Loc0: 0.026, 0.029, 0.030, 0.029, 0.036 Loc1: 0.119, 0.000, 0.000, 0.000, 0.000
10
```

6.2.2 Example of a Nonrandom Mating in a Continuous Habitat

One of the most common factors causing nonrandom mating is geographic location. If a habitat is fragmented into small demes, random mating can be assumed within each deme and exchange of genotypes between demes can be achieved through migration. The genetic distance between these demes is determined by the relative strength of genetic drift and migration. This model has been simulated in a few occasions in this book.

However, in a continuous habitat where no obvious fragmentation seems to prevent individual movement, genetic drift can happen locally due to preferential mating with a spouse in geographic vicinity. Because exchange of genotypes happens locally among geographically close individuals, genetic information flows gradually over the entire habitat and leads to a pattern of local similarity and increasing differentiation with distance. Such a model can be simulated by using stepping-stone migration models in one or two dimensions if we assume random mating within clusters of individuals [5] or by using nonrandom mating schemes that choose pairs of parents according to their geographic locations.

■ EXAMPLE 6.5

This example defines a mating scheme `VicinityMating` that implements mating in a one-dimensional continuous habitat. Instead of using a global generating function (see Example 6.4), this example derives a mating scheme from class `HomoMating`. This mating scheme uses a Python parents chooser that calls a member function `_chooseParents` and uses a PyTagger to set geographic location of offspring in addition to user-provided during-mating operators.

In order to select parents according to geographic vicinity, function _chooseParents first sorts individuals by values at an information field. It then chooses a parent randomly, finds its location p_1, identifies all parents with opposite sex within the range of $(p_1 - v, p_1 + v)$, and randomly chooses a parent from them. If an individual is too far away from others, he or she will not be able to find a spouse. The geographic location of an offspring will be determined by a normal distribution with a mean that is the average of parental location and a specified variable.

To test this mating scheme, we created a population of 2000 individuals. An information field x is used to record geographic location of each individual. Although the habitat is continuous, we define eight VSPs by this information field ($x < 1$, $1 \leq x < 2$, ..., $6 \leq x < 7$, and $x > 7$). Individuals in this population are scattered uniformly along $0 < x < 8$. We initialize individuals with $4 \leq x < 5$ with a mutant allele with frequency 0.4 and uses a VicinityMating mating scheme to mate parents randomly with an individual who is within 1 unit left or right to it. As we can see from the output of this example, the mutant spread slowly from the center to all geographic regions.

SOURCE CODE 6.5 A Mating Scheme with Continuous Habitat

```
>>> import simuPOP as sim
>>> from random import randint, uniform, normalvariate
>>> class VicinityMating(sim.HomoMating):
...     '''A homogeneous mating scheme that choose parents that are close to
...     each other according to values at an information field.'''
...     def __init__(self, locationField='x', varOfLocation=1, vicinity=1,
...         numOffspring=1, sexMode=sim.RANDOM_SEX, ops=sim.MendelianGenoTransmitter(),
...         subPopSize=[], subPops=sim.ALL_AVAIL, weight=0):
...         '''Creates a random mating scheme that selects a parent randomly and
...         another random parent who is in vivinity with him/her, namely with
...         location that is within [x-v, x+v] where x is the location of the first
...         parent, and v is specified by parameter vicinity. For each offspring,
...         its location is set according to a normal distribution with a mean that
...         is the average of parental locations, and a variance varOfLocation.
...         '''
...         self.field = locationField
...         self.vicinity = vicinity
...         self.varOfLocation = varOfLocation
...         if hasattr(ops, '__iter__'): # if a sequence is given
...             # WithArgs is needed because field name is a variable.
...             allOps = ops + [sim.PyTagger(sim.WithArgs(self._passLocation, [self.field]))]
...         else:
...             allOps = [ops, sim.PyTagger(sim.WithArgs(self._passLocation, [self.field]))]
...         sim.HomoMating.__init__(self,
...             chooser = sim.PyParentsChooser(self._chooseParents),
```

```
...             generator = sim.OffspringGenerator(allOps, numOffspring, sexMode),
...             subPopSize = subPopSize,
...             subPops = subPops,
...             weight = weight)
...
...     def _passLocation(self, field):
...         return normalvariate((field[0]+field[1])/2, self.varOfLocation)
...
...     def _chooseParents(self, pop, subPop):
...         # sort individuals according to location
...         pop.sortIndividuals(self.field)
...         while True:
...             # select the first parent
...             p1 = randint(0, pop.subPopSize(subPop) - 1)
...             x1 = pop.individual(p1).info(self.field)
...             s1 = pop.individual(p1).sex()
...             # find all invividuals with opposite sex within vivinity of p1
...             inds = []
...             p = p1 + 1
...             while p < pop.subPopSize(subPop) and \
...                 pop.individual(p, subPop).info(self.field) < x1 + self.vicinity:
...                 if pop.individual(p, subPop).sex() != s1:
...                     inds.append(p)
...                 p += 1
...             p = p1 - 1
...             while p >= 0 and \
...                 pop.individual(p, subPop).info(self.field) > x1 - self.vicinity:
...                 if pop.individual(p, subPop).sex() != s1:
...                     inds.append(p)
...                 p -= 1
...             # if no one is invicinity, find another pair
...             if len(inds) == 0:
...                 continue
...             # choose another parent
...             p2 = inds[randint(0, len(inds) -1)]
...             # return indexes of both parents
...             if s1 == sim.MALE:
...                 yield p1, p2
...             else:
...                 yield p2, p1
...
>>> pop = sim.Population(size=2000, loci=1, infoFields='x')
>>> # define VSPs x<1, 1<=x<2, 2<=x<3, 3<=x<4, ...
>>> pop.setVirtualSplitter(sim.InfoSplitter(field='x', cutoff=range(1, 8)))
>>> pop.evolve(
...     initOps=[
...         sim.InitSex(),
...         sim.InitInfo(lambda : uniform(0, 8), infoFields='x'),
...         # only individuals in the middle range has certain genotype
...         sim.InitGenotype(freq=[0.6, 0.4], subPops=[(0, 4)]),
...     ],
...     matingScheme=VicinityMating(locationField='x', vicinity=1, varOfLocation=0.5),
...     postOps=[
...         sim.Stat(alleleFreq=0, subPops=[(0, sim.ALL_AVAIL)], vars='alleleFreq_sp'),
...         sim.PyEval(r"'%.3f ' % alleleFreq[0][1]", subPops=[(0, sim.ALL_AVAIL)]),
...         sim.PyOutput('\n'),
...     ],
...     gen = 10
```

```
... )
0.000 0.000 0.004 0.092 0.229 0.090 0.004 0.000
0.000 0.000 0.011 0.085 0.186 0.079 0.022 0.000
0.002 0.005 0.026 0.078 0.139 0.073 0.028 0.000
0.002 0.007 0.035 0.082 0.106 0.074 0.046 0.016
0.008 0.003 0.039 0.079 0.104 0.080 0.057 0.010
0.006 0.013 0.031 0.106 0.114 0.074 0.086 0.027
0.006 0.007 0.060 0.081 0.093 0.103 0.086 0.045
0.009 0.011 0.050 0.091 0.089 0.064 0.088 0.047
0.006 0.019 0.062 0.072 0.076 0.064 0.066 0.042
0.004 0.021 0.048 0.071 0.075 0.073 0.061 0.051
10
>>>
```

6.3 HETEROGENEOUS MATING SCHEMES

Individuals in a population do not have to share the same mating pattern. Different mating patterns might exist for individuals with different or even the same properties. For example, individuals with different social status in human populations exhibit different mating patterns, and certain plant species exhibit both self- and cross-fertilizations in a more or less random way. These mating schemes can be simulated using heterogeneous mating schemes in simuPOP.

■ EXAMPLE 6.6

A heterogeneous mating scheme (`HeteroMating`) accepts a list of homogeneous mating schemes and apply them to different subpopulations or virtual subpopulations. For example, a heterogeneous mating scheme

```
HeteroMating([
    RandomMating(numOffspring=1, subPops=0),
    RandomMating(numOffspring=2, subPops=1)
])
```

applies two random mating schemes with different parameters to subpopulations 0 and 1 and

```
HeteroMating([
    SelfMating(subPops=0),
    RandomMating(subPops=[1,2])
])
```

applies a self-fertilization mating scheme to the first subpopulation and a random mating scheme to other subpopulations.

The real power of heterogeneous mating schemes lies on their ability to apply different mating schemes to different virtual subpopulations. If

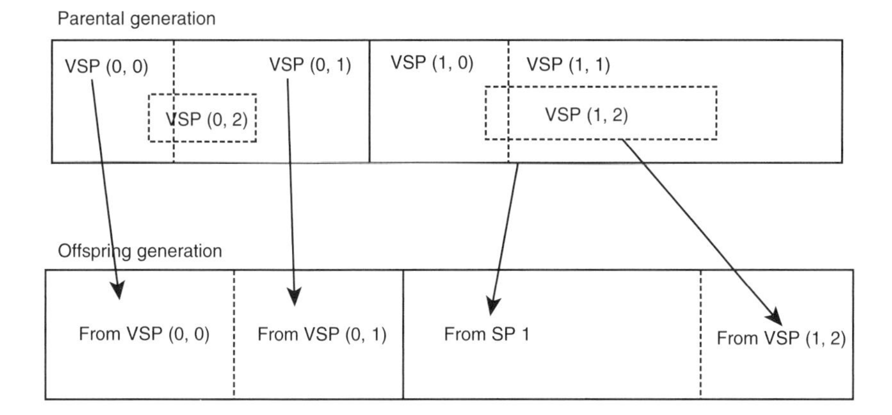

FIGURE 6.1 Illustration of a heterogeneous mating scheme. A heterogeneous mating scheme that applies four homogeneous mating schemes to virtual subpopulations (0,0), (0,1), subpopulation 1, and virtual subpopulation (1,2). The first and last two mating scheme populations are the first and second subpopulations of the offspring population, respectively.

a mating scheme is applied to a virtual parental subpopulation, parents will be selected from that VSP. If multiple mating schemes are applied to the same subpopulation, each of them populates only part of the offspring subpopulation. This is illustrated in Figure 6.1 where four homogeneous mating schemes are applied to different virtual and nonvirtual subpopulations and populate two subpopulations of the offspring population. The first two mating schemes choose parents from two nonoverlapping VSPs, so parents in this subpopulation will use one of the two mating schemes according to the VSPs they belong to. The third mating scheme is applied to the whole subpopulation and the fourth mating scheme is applied to a VSP (1,2). Because parents in VSP (1,2) also belong to subpopulation 1, they might be involved in two different mating schemes.

Due to different microenvironmental factors, plants in the same population may exercise both self- and cross-fertilizations. Because offspring resulting from both self- and cross-fertilizations can themselves reproduce with self- or cross-fertilizations, dividing a population into self- and cross-mating subpopulations cannot be used to simulate this evolutionary process. This example demonstrates how to simulate partial self-fertilization using a heterogeneous mating scheme with two mating schemes. This example applies a `SelfMating` mating scheme and a `RandomMating` mating scheme to two VSPs of the first subpopulation defined by proportions of individuals. The VSPs are defined by a `ProportionSplitter` so that 20% of individuals go through self-mating and 80% of individuals go through random mating. The proportions of offspring produced by these two mating schemes are 80% and 20%, respectively, because, by default, the number of offspring each mating scheme produces is proportional to parental (virtual) subpopulation sizes. As a comparison, we apply the random mating scheme to a second subpopulation and compare the observed homozygosity between these two subpopulations. Because selffertilization tend to increase homozygosity, the first subpopulation exhibits higher homozygosity than the second one.

SOURCE CODE 6.6 Use of a Heterogeneous Mating scheme to Simulate Partial Self-Fertilization

```
>>> import simuPOP as sim
>>> pop = sim.Population(size=[10000, 10000], loci=1)
>>> pop.setVirtualSplitter(sim.ProportionSplitter([0.8, 0.2]))
>>> pop.evolve(
...     initOps=[
...         sim.InitSex(),
...         sim.InitGenotype(freq=[0.5, 0.5])
...     ],
...     preOps=[
...         sim.Stat(homoFreq=0, subPops=[0,1], vars='homoFreq_sp'),
...         sim.PyEval(r"'(%.2f, %.2f)\n' % (subPop[0]['homoFreq'][0], "
...             "subPop[1]['homoFreq'][0])"),
...     ],
...     matingScheme=sim.HeteroMating(matingSchemes=[
...         sim.RandomMating(subPops=[(0, 0), 1]),
...         sim.SelfMating(subPops=[(0, 1)]),
...     ]),
...     gen = 3
... )
(0.50, 0.50)
(0.56, 0.49)
(0.55, 0.49)
3
```

Example 6.3 implements a positive assortative mating scheme by choosing parents sequentially. Because individuals are sorted by a quantitative trait, this mating scheme effectively matches parents with high desirability (trait D) with spouses with high desirability, and parents with low desirability with spouses with low desirability. If a less stringent assortative mating scheme is used, namely, a parent with high desirability can mate with spouses with certain range of desirability, a heterogeneous mating scheme can be used. If we divide trait D into three categories as high, medium, and low desirability using two cutoff points, we can force all mating events to happen within their own desirability groups. It is also easy to add certain level of noise to this mating scheme by allowing mating across desirability groups (random mating across the whole population) for certain proportion of parents.

6.3.1 Simulation of Population Admixture

Population structure has been known to cause spurious associations in case–control association studies [6], so several statistical methods have been developed to reduce the impact of population structure on GWA studies [7–9]. On the other hand, population admixture causes long-range admixture LD that could be used to map diseases in admixed populations [10, 11]. Although simulations have been used to evaluate the performance of these statistical methods, they have not been complex enough to challenge the statistical methods under realistic situations [12]. For example, Pfaff et al. [13] broke existing LD of the founder populations from HapMap samples by sampling alleles instead of haplotypes so that only admixed LD existed in the simulated sample.

We aimed to simulate realistically admixed populations by mixing populations simulated from the HapMap populations. We extracted 5000 markers with minor allele frequency greater than or equal to 0.05 from chromosome 2 (chr2:50002476-60382263) using 170 and 143 independent individuals from phase 2 of the HapMap Japanese in Tokyo, Japan and Han Chinese in Beijing, China (JPT + CHB) and Maasai in Kinyawa, Kenya (MKK) populations. The two populations were expanded to a total size of 50,000 individuals. A low-level migration rate of 0.0001 was applied to keep the genetic distance between these two populations around its original level of 0.11 (measured using FST) [14].

To control the levels of true LD and admixture LD, we mixed large populations to avoid elevated LD caused by a founder effect. We mixed these two populations using a continuous gene flow model where 5% of individuals from MKK population migrated to JPT + CHB population for 10 generations [15]. At the beginning of population mixing, we assigned an ancestral value of 0 to individuals from JPT + CHB population and a value of 1 for individuals from MKK population. During the admixture process, the offspring ancestral values were recorded as the mean of parental ancestral values. We used a positive assortative mating scheme to mix individuals because migrants usually do not mate randomly with natives during a real-world admixture process, and individuals would be efficiently mixed and have similar ancestral values only after a few generations if the standard Wright–Fisher random mating process were used to mix parents regardless of their ethnicity. More specifically, we divided individuals into two groups according to their ancestral value, one with ancestral values greater than or equal to 0.5 and another with ancestral values less than 0.5. During the offspring population generation, 80% of all mating events happened within these two groups and the rest of the mating events happened with parents chosen randomly from the whole population [16]. This process slowed down the admixture process and resulted in a distribution of individual ancestry values that is closer to that of real populations, such as the mixture distribution of European ancestry among all African Americans [17]. Example 6.7 demonstrates how to implement such an admixing process using a heterogeneous mating scheme.

■ EXAMPLE 6.7

We use scripts `loadHapMap3.py` and `selectMarkers.py` to prepare an initial population and use function `simuGWAS` defined in Example 4.6 to create a large population with two subpopulations (code not listed).

We add two information fields `ancestry` and `migrate_to` to this population and initialize individuals in the first population with ancestry value 0 and individuals in the second population with ancestry value 1. Individuals are grouped into two virtual subpopulations, one with `ancestry < 0.5` and the other with `ancestry >= 0.5`.

A migrator is used at the beginning of each generation to migrate individuals from the second subpopulation to the first at a rate of 5% per generation. This is a directional migration because no one migrates from the first to the second subpopulation.

We use a heterogeneous mating scheme with three mating schemes to perform random mating in the second subpopulation and a nonrandom mating in the first subpopulation. Because the second and third mating schemes are applied to virtual subpopulations `(0,0)` and `(0,1)` in the first subpopulation, the first mating schemes will produce all offspring in the second subpopulation.

The number of offspring produced by these three mating schemes is controlled by parameter `weight`. Briefly speaking,

- A negative weight is considered proportional to the source parental (virtual) subpopulation size. Negative weights are handled before positive or zero weights.
- If all remaining weights are zero, the number of offspring each mating scheme produces is proportional to its parental (virtual) subpopulation sizes.
- If there is any positive weight, the number of offspring each mating scheme produces is proportional to the assigned weight. Mat-

ing schemes with zero weight will not produce any offspring in this case.

Therefore, the first mating scheme will be applied to the whole subpopulation and will produce 20% of the offspring, the second and third mating schemes will be applied to two virtual subpopulations and will produce 80% of offspring.

In addition to a Mendelian genotype transmitter that transmit parental genotypes to offspring, all these mating schemes apply a `InheritTagger` during mating, which passes the average of parental ancestral values to their offspring. This value therefore records the true ancestry proportion for all individuals.

SOURCE CODE 6.7 Simulating an Admixed Population with Recorded Ancestral Values

```
import simuOpt
simuOpt.setOptions(gui=False, alleleType='binary')
import simuPOP as sim
pop.addInfoFields(['ancestry', 'migrate_to'])
# initialize ancestry
sim.initInfo(pop, [0]*pop.subPopSize(0) + [1]*pop.subPopSize(1),
    infoFields='ancestry')
# define two virtual subpopulations by ancestry value
pop.setVirtualSplitter(sim.InfoSplitter(field='ancestry', cutoff = [0.5]))
transmitters=[
    sim.MendelianGenoTransmitter(),
    sim.InheritTagger(mode=sim.MEAN, infoFields='ancestry')]
pop.evolve(
    initOps=sim.InitSex(),
    preOps=sim.Migrator(rate=[
        [0., 0], [0.05, 0]]),
    matingScheme=sim.HeteroMating(
        matingSchemes=[
            sim.RandomMating(ops=transmitters),
            sim.RandomMating(subPops=[(0,0)], weight=-0.80, ops=transmitters),
            sim.RandomMating(subPops=[(0,1)], weight=-0.80, ops=transmitters)
        ],
    ),
    gen=10,
)
# remove the second subpop
pop.removeSubPops(1)
```

We applied to the population a penetrance model in which individuals' probability of being affected equals 0.05 + ancestry/6, where ancestry is the individual's MKK ancestry value. Individuals with higher MKK ancestry values were more susceptible to this disease, but none of the 5000 markers caused the disease directly. Because there were enough affected individuals in the simulated population, we drew 500 cases and 500 controls directly from the simulated population. We used the STRUCTURE program to estimate the ancestry values of cases and controls from their genotypes [7] and plotted the estimated MKK ancestries against the recorded ancestry values for each individual (Figure 6.2a and b). Because individuals with high MKK ancestry values are more likely to be affected, cases on average had higher MKK ancestry values than controls.

In order to demonstrate the impact of population structure on association analysis, we applied allele-based χ^2 tests and structured association tests proposed by Pritchard et al. [18] to detect the association between disease status and 2000 simulated markers (Figure 6.2c and d). Although the disease is not directly caused by any of the simulated markers, a large number of spurious associations were detected by the χ^2 tests. In contrast, the structured association tests estimated individual ancestry values to control the impact of population structure and successfully removed most spurious associations.

6.4 SIMULATION OF AGE-STRUCTURED POPULATIONS

The Wright–Fisher model uses a nonoverlapping generation model that replaces a parental generation with its offspring generation once the offspring generation is created. Because parents in a nonoverlapping generation model do not stay in the same population as their offspring, such a model reflects life cycles of semelparous populations such as annual plants (e.g., all grain crops), salmon, or bamboos in which parents die shortly after reproduction. It is certainly not realistic for iteroparous populations such as the human populations in which parents have many reproductive cycles over the course of its lifetime.

A nonoverlapping generation model can be used to approximate the evolution of human populations because humans usually choose one spouse from his or her age group (generation), transmit their genotypes to the offspring generation, and make no further contribution to the evolutionary process. If we look at the long-term impact of genetic and demographic factors on the evolution of human populations, we can synchronize genealogies to form reproducing generations separately by roughly 20 years and simulate such a population using a nonoverlapping evolutionary model.

However, if we are studying human populations at a finer scale, it becomes difficult to ignore the fact that only a small percentage of humans reproduces at any given time, and that people with different ages stay in the same population. For example, if we are going to simulate a population

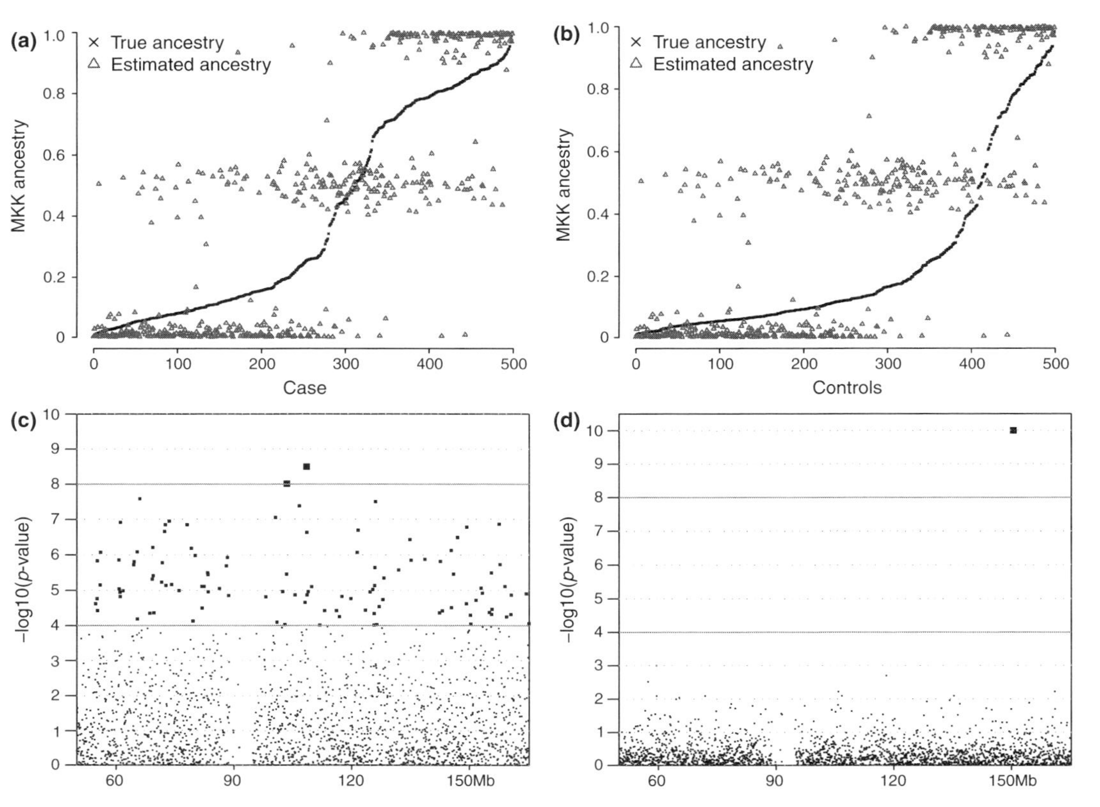

FIGURE 6.2 Gene mapping in an admixtured population. Ancestry values and p-values of association tests. The top figures plot recorded and estimated MKK ancestry values of 500 cases (a) and 500 controls (b). Individuals are sorted by their true MKK ancestry values. The bottom figures plot the negative of the base 10 logarithm of p-values of allele-based χ^2 tests (c) and structured association tests (d) between 500 cases and 500 controls at 2000 markers.

with lung cancer, it is necessary to include individuals with different ages because age has strong impact on this disease.

6.4.1 Simulation of Age-Structured Populations

An age-structured population consists of individuals with different ages. The evolution of an age-structured population differs from a standard Wright–Fisher models in the following ways:

- Instead of evolving by generation, an age-structured population usually evolves by year or by stage (age-group).
- Instead of surviving for a single generation, individuals in an age-structured population have a life history and will typically stay in the population for several years.
- Mating in age-structured population is usually not random. A typical scenario is that only individuals within certain age range can mate and produce offspring.
- Parents usually do not die immediately after the production of offspring so it is likely that parents and offspring will coexist in the same population.

Although it is infeasible to directly simulate the evolution of age-structured populations using a discrete generation model, such evolutionary scenarios can be imitated through the use of nonrandom mating schemes under a discrete generation simulation framework. The key here is to use a mating scheme that copies individuals from parental to offspring generation and in the mean time produces offspring from selected parents. Example 6.8 demonstrates how to achieve this using a heterogeneous mating scheme of simuPOP.

The Wright–Fisher model uses a nonoverlapping generation model without age structure, so parents and offspring cannot stay at the same population. In order to simulate the evolution of an age-structured population, it is necessary to copy eligible parents from parental to offspring populations. More specifically, you will need to do the following:

■ EXAMPLE 6.8

Example 6.8 gives an example of the evolution of age-structured population. It defines an information field `age` and uses it to store age of all individuals. The concept of age here can be extended to stage. For example, an individual can grow from newborn to junior, adult, and senior, instead of growing year by year. In addition to `age`, this population also defines information fields `ind_id`, `father_id` and `mother_id` in order to track parental relationship in the age-structured population.

In order to determine who should be copied to the offspring population and who are eligible to mate, this example uses a `InfoSplitter` to divide the population into four virtual subpopulations by age. These virtual subpopulations include junior (age <20), adult ($20 \leq$ age <50), senior ($50 \leq$ age <75), and people of age ≥ 75. We assume that people of age <75 will survive to the next generation (year) and people with age ≥ 75 will die and be removed from the population. We assume random mating among adult individuals, so individuals with $20 \leq$ age <50 will mate randomly with each other and produce offspring.

The simulation starts from the creation of a population of 10,000 individuals. Before evolution, individuals in this population are initialized with random sex, age, and genotype and are assigned a unique ID. At the beginning of each generation, individual age is increased by 1, so all individuals will stay for exactly 75 years (generations) except for those with a nonzero initial age at the beginning of the evolution.

The nonrandom mating scheme described above is implemented using a heterogeneous mating scheme that consists of two mating schemes:

- A `CloneMating` mating scheme copies all parents with age less than 75 to the offspring population. A -1 weight is used, so everyone is copied once and only once. This mating scheme uses a `CloneGenoTransmitter`, which clones not only the genotype but also sex and all information fields. IDs of parents will therefore be kept so that individuals will be able to locate their parents even if they are now in the same population. Note that individuals with duplicate IDs will occur if you decide to store some parental generations during simulation.
- A `RandomMating` mating scheme is used to produce offspring from parents in the reproducing age (`subPops=[(0,1)]`). Three operators are used to assign a unique ID, record parental IDs, and transmit genotype for each offspring. This mating scheme produces 1, 2, or 3 offspring at each mating event so that there will be nuclear families with varying sizes in the resulting population.

The population is evolved for 200 generations. Just to verify that we can identify parents and their offspring from the resulting population, we use function `drawNuclearFamilySample` to draw a family with two or three offspring from the resulting population. Individuals in the selected family is displayed using a dump function.

SOURCE CODE 6.8 Example of the Evolution of Age-Structured Population

```
>>> import simuPOP as sim
>>> from random import randint
>>> pop = sim.Population(10000, loci=1,
...     infoFields=['age', 'ind_id', 'father_id', 'mother_id'])
>>> pop.setVirtualSplitter(sim.InfoSplitter(field='age', cutoff=[20, 50, 75]))
>>> pop.evolve(
...     initOps=[
...         sim.InitSex(),
...         # random assign age
...         sim.InitInfo(lambda: randint(0, 74), infoFields='age'),
...         sim.InitGenotype(freq=[0.5, 0.5]),
...         # assign an unique ID to everyone.
...         sim.IdTagger(),
...     ],
...     # increase the age of everyone by 1 before mating.
...     preOps=sim.InfoExec('age += 1'),
...     matingScheme=sim.HeteroMating([
...         # all individuals with age < 75 will be kept. Note that
...         # CloneMating will keep individual sex, affection status and all
...         # information fields (by default).
...         sim.CloneMating(subPops=[(0,0), (0,1), (0,2)], weight=-1),
...         # only individuals with age between 20 and 50 will mate and produce
...         # offspring. The age of offspring will be zero.
...         sim.RandomMating(ops=[
...             sim.IdTagger(),                     # give new born an ID
...             sim.PedigreeTagger(),               # track parents of each individual
...             sim.MendelianGenoTransmitter()],    # transmit genotype
...             numOffspring=(sim.UNIFORM_DISTRIBUTION, 1, 3),
...             subPops=[(0,1)])
...     ]),
...     gen = 200
... )
200
>>>
>>> from simuPOP import sampling
>>> sample = sampling.drawNuclearFamilySample(pop, families=1, numOffspring=(2,3))
>>> sim.dump(sample, structure=False)
SubPopulation 0 (), 4 Individuals:
   0: FU 0 | 0 |  14 34676 30053 28957
   1: MU 1 | 0 |  49 30053 27189 25057
   2: FU 0 | 0 |  14 34677 30053 28957
   3: FU 0 | 0 |  58 28957 24865 24284
```

6.4.2 A Hypothetical Disease Model

We use a proportional hazards model to model the risks of lung caner. We assume that individuals who do not suffer from any genetic or environmental risk factor has the lowest risk of getting lung cancer and model their risks with a hazard function

$$h_0(t) = \beta_0 \exp(a_0(t-18)),$$

where t is age in years and β_0 and α_0 are sex-dependent parameters. $h(t)$ is assumed to be zero if $t \leq 18$ because teenagers rarely get lung cancer.

The hazard functions for individuals with certain genetic or environmental risk factors are assumed to be $h(t) = \beta \exp(a_0(t-18))$ which is proportional $h_0(t)$. Furthermore, we assume that all these risk factors influence β multiplicatively. For example, if a male smoker carries a disease allele at locus G_2, his risk of getting lung cancer at the age t will be modeled by a hazard function

$$h(t) = r_{\mathrm{MS}} r_{G1} \beta_0 \exp(a_0(t-18)),$$

where r_{MS} and r_{G1} are relative risks for male smoker and for G_1 carriers.

We assume that lung cancer is caused by seven interacting genetic and environmental risk factors. More specifically, we assume that there is a gene G_1 that contributes to smoking (e.g., through nicotine addition), which in turn influences the risk of COPD and lung cancer. The risk of COPD is also influenced by a gene G_4 and if someone has COPD, he or she has increased risk of getting lung cancer. Two other genes G_2 and G_3 also contribute to the risk of lung cancer, but G_3 only contributes to the risk of lung cancer of smokers. In addition, a random genetic factor exists and influences the risk of lung cancer with a relative risk directly. Such a model is illustrated in Figure 6.3.

For each individual, we determine whether or not he or she will smoke and simulate the age at which he or she gets COPD and lung cancer according to his or her genotype, smoking status, and a random environmental factor. Because

$$h(t) = \frac{dF(t)/dt}{1-F(t)},$$

we calculate

$$F(t) = 1 - \exp\left(-\int_0^t h(\theta)\, d\theta\right)$$

and determine the age of onset using

$$t = F^{-1}(u), \tag{6.1}$$

where u follows a uniform distribution between 0 and 1.

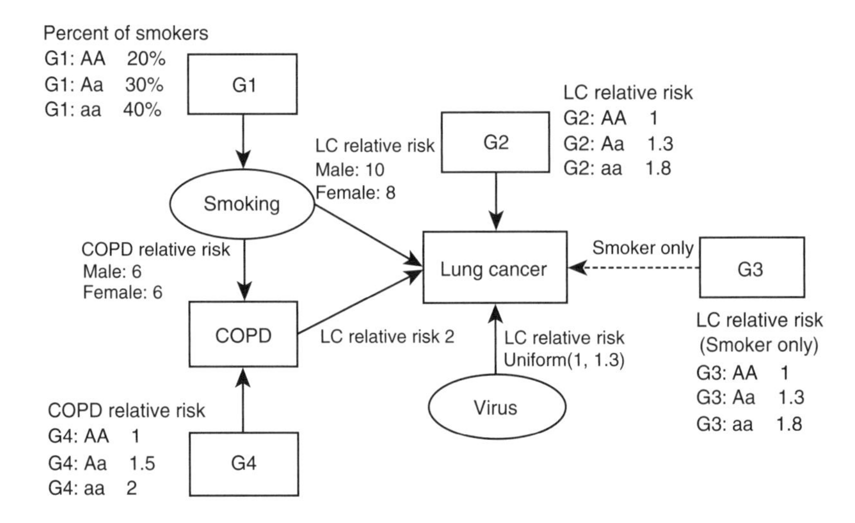

FIGURE 6.3 A hypothetical lung cancer risk model. A hypothetical lung cancer risk model with four genetic factors, smoking, chronic obstructive pulmonary disease (COPD), and a random environmental factor.

For each individual, we simulate his or her smoking status, age at which he or she is affected with lung cancer, and age of death according to genotypes at four loci and a random effect. For example, if a male individual has genotype Aa at locus G_1, he will have 30% of chance to become a smoker. The age at which he is affected with COPD will be calculated from Equation 6.1 according to a personal hazard function determined by his smoking status and his genotype at locus G_4. This age will be compared with his age of death (a random number drawn from a uniform distribution from age 60 to 80) to determine whether or not he will be affected with COPD before his end of life. These information are then combined with his genotypes at G_2 and G_3 and a random effect to determine the age at which he will be affected with lung cancer. We assume that all lung cancer patients will die after 6 years and adjust the age of death of lung cancer patients accordingly.

■ EXAMPLE 6.9

Because genotype of individuals are fixed when he or she is born, we determine an individual's smoking status, age of onset of lung cancer, and age of death and store these information in information fields `smoking`, `age_LC`, and `age_death`. This procedure is implemented in function `initialize` of the `LC_model` class. This function will be used in Example 6.10

to initialize smoking status, age of onset, and death for individuals in the starting population and for all offspring when they are born.

SOURCE CODE 6.9 Implementation of the Lung cancer Disease Model

```
import simuPOP as sim
from random import uniform, randint
from math import exp

class LC_model:
    def __init__(self, LC_beta0_male, LC_beta0_female, LC_a0, COPD_beta0,
        COPD_a0, G1_smoking_rate, rr_G2, rr_G3, rr_G4, rr_random,
            rr_LC_maleSmoker, rr_LC_femaleSmoker, rr_COPD_smoker,
            rr_LC_COPD):
        '''LC model with parameters for different relative risks
        A disease model is responsible for updating age_LC, age_death, and
        smoking. It can use other information fields to its own use.
        '''
        self.LC_beta0_male, self.LC_beta0_female, self.LC_a0, self.COPD_beta0, \
        self.COPD_a0, self.G1_smoking_rate, self.rr_G2, self.rr_G3, self.rr_G4, \
        self.rr_random, self.rr_LC_maleSmoker, self.rr_LC_femaleSmoker,  \
        self.rr_COPD_smoker, self.rr_LC_COPD = \
            LC_beta0_male, LC_beta0_female, LC_a0, COPD_beta0, \
            COPD_a0, G1_smoking_rate, rr_G2, rr_G3, rr_G4, rr_random, \
            rr_LC_maleSmoker, rr_LC_femaleSmoker, rr_COPD_smoker, rr_LC_COPD


    def _cdf(self, t, beta, a):
        # use a stepwise function to approximate integration
        ch = sum([beta*exp((i-18)*a) for i in range(18, t+1)])
        v = 1 - exp(-ch)
        return v

    def _ageOfOnset(self, beta, a):
        '''Calculate age of onset with a given beta and a. This is only used to initialize
        a population without considering cessation and other factors.
        '''
        u = uniform(0, 1)
        # a bisection method will be more efficient..
        for age in range(80):
            if self._cdf(age, beta, a) > u:
                return age
        return 100

    def initialize(self, ind):
        '''Determines the risk of LC for passed offividual. The return value is the
        multiple of a base hazard function.
        '''
        geno = [ind.allele(x,0) + ind.allele(x,1) for x in range(4)]
        # smoking, determined by G1
        ind.smoking = uniform(0,1) < self.G1_smoking_rate[geno[0]]
        # original age of death
        ind.age_death = randint(60, 80)
        # coefficient for LC and COPD
        LC_beta = self.LC_beta0_male if ind.sex() == sim.MALE else self.LC_beta0_female
        COPD_beta = self.COPD_beta0
```

```
        # smoking
        if ind.smoking:
            if ind.sex() == sim.MALE:
                LC_beta *= self.rr_LC_maleSmoker
            else:
                LC_beta *= self.rr_LC_femaleSmoker
            COPD_beta *= self.rr_COPD_smoker
        # G2
        LC_beta *= self.rr_G2[geno[1]]
        # G3
        if ind.smoking:
            LC_beta *= self.rr_G3[geno[2]]
        # G4
        COPD_beta *= self.rr_G4[geno[3]]
        if self._ageOfOnset(COPD_beta, self.COPD_a0) < ind.age_death:
            LC_beta *= self.rr_LC_COPD
        # random factor
        LC_beta *= 1 + self.rr_random * uniform(0,1)
        # LC?
        ind.age_LC = self._ageOfOnset(LC_beta, self.LC_a0)
        # adjust age of death if someone will get LC
        if ind.age_death < ind.age_LC + 6:
            ind.age_death = ind.age_LC + 6
        return ind.age <= ind.age_death

    def updateStatus(self, pop):
        # required by the evolutionary process but this disease model
        # currently does not need to update individual status dynamically.
        return True
```

We determine the age of onset of each individual when he or she is born, but this simulation model can be easily extended so that individual susceptibility to lung cancer can be variable in response to events such as cessation. In this case, $h(t)$ of individuals can vary year by year and individual affection status should be calculated as

$$\begin{aligned} G(t) &= P(t \le T < t+1 \mid T \ge t) \\ &= \frac{F(t+1) - F(t)}{1 - F(t)} \\ &= 1 - \exp\left(-\int_t^{t+1} h(\theta)\, d\theta\right). \end{aligned}$$

If we assume $h(\theta)$ is roughly constant between t and $t+1$,

$$G(t) \sim 1 - \exp(-h(t)) \sim h(t) + \frac{1}{2} h(t)^2,$$

so we can use $h(t)$ to approximate $G(t)$ at each year.

6.4.3 Evolution of an Age-Structured Population with Lung Cancer

Example 6.10 simulates an evolutionary scenario where an age-structured population with lung cancer is evolved year by year. This simulation scenario can be used to observe the impact of environmental factors on the incidence of lung cancer (or other complex human diseases) and to generate realistic samples for the development of efficient gene mapping methods for the detection of gene–environment interaction for such a disease. For simplicity, this example simulates a stable environmental factor (smoking pattern), so the disease prevalence will stabilize over time.

The value of such an evolutionary process lies in the fact that it can closely mimic real human populations with complex population structure and changing environmental factors. For example, within the United States, historical changes in tobacco products and in patterns of usage over the last century have dramatically influenced the number of people that smoke [19]. These patterns illustrate the wave-like uptake of cigarette smoking by sequential generations of Americans. The incidence of LC and the corresponding mortality rate in the U.S. population parallel these tobacco smoking patterns, with a time lag of about 30 years [19]. By simulating such a pattern of smoking, we can observe the dynamic of disease prevalence as a result of changing smoking behavior. Such a simulation can be used to validate individual disease models, and once a disease model has demonstrated realistic individual, familial, and population properties, it can be used to predict the population prevalence of lung cancer when certain cancer prevention strategies are used to change population smoking patterns.

■ EXAMPLE 6.10

This example defines a function `LC_evolve` that evolves an age-structured population with lung cancer using a disease model described in Example 6.9. In this function, a population is first created with four unlinked loci and five information fields `age`, `smoking`, `age_death`, `age_LC`, and `LC`. Among these information fields, `age` is the age of individual that will be updated at each year (generation), `smoking`, `age_death`, and `age_LC` are determined by function `initialize` of a disease model, and `LC` is determined by whether or not individual `age` is greater than `age_LC`.

A virtual splitter `CombinedSplitter` is assigned to the population to categorize individuals by age, sex, and smoking status (lines 10–17). This splitter consists of two `InfoSplitter` and a `SexSplitter`, which define virtual subpopulations 0, 1, and 2 with individuals of $\text{age} < 20, 20 \leq \text{age} < 40$, and $\text{age} \geq 40$, respectively; virtual subpopulations 3 and 4 with male and female individuals, respectively; and virtual subpopulations 5, 6, and 7 according to smoking status. In order to avoid confusion, meaningful names are given to each virtual subpopulation and are used to refer to these VSPs later.

Before evolution, seven operators are used to initialize individual sex, age, genotype at four loci, and information fields (lines 20–23). Instead of assigning random ages as we have done in Example 6.8, this example uses a sequence 0, 1, ..., 74 to initialize ages of all individuals in the population sequentially. After individual sex, age, and genotypes are initialized, function `initialize` of the passed disease model object is used to determine `smoking`, `age_death`, and `age_LC` of each individual. The function will return `False` when the age of an individual is greater than his or her age of death, which will effectively remove these individuals from the starting population. Note that the `PyOperator` applies the function to all individuals one by one because this function is defined with parameter `ind`. If a parameter `pop` is used, the function would be applied to the whole population.

The population is then evolved year by years, marked by increased age of all individuals (line 25). The increase of age will cause the death (removal from the population) of some individuals if their age is greater than their age of death (line 26).

A heterogeneous mating scheme is then used to evolve the population using two mating schemes. The first mating scheme copies all individuals from the parental to the offspring population, and the second mating scheme uses a `RandomMating` scheme to produce individuals from parents between the age of 20 and 40. The offspring will have a random sex, an initial age of zero, and will be initialized by function `initialize` of the disease model object. This mating scheme produces exactly `popSize/75` offspring because it uses a demographic function

```
lambda pop: pop.popSize() + popSize/75
```

and the first `pop.popSize()` individuals will be produced by the clone mating scheme because of use of a `-1` weight.

At the end of each year, the affection status of each individual is determined by comparing each individual's age with his or her age of onset of lung cancer. This is achieved by evaluating statement `LC = age >= age_LC` for each individual using operator `InfoExec`. The population prevalence of lung cancer is then calculated at each generation by calculating the mean of information field LC in all virtual subpopulations.

SOURCE CODE 6.10 Evolution of Lung Cancer

```
import simuPOP as sim

from ch7_LC_model import LC_model

def LC_evolve(popSize, alleleFreq, diseaseModel):
    '''
    '''
    pop = sim.Population(size=popSize, loci=[1]*len(alleleFreq),
        infoFields = ['age', 'smoking', 'age_death', 'age_LC', 'LC'])
    pop.setVirtualSplitter(sim.CombinedSplitter(splitters=[
        sim.InfoSplitter(field='age', cutoff=[20, 40],
            names=['youngster', 'adult', 'senior']),
        sim.SexSplitter(),
        sim.InfoSplitter(field='smoking', values=[0, 1, 2],
            names=['nonSmoker', 'smoker', 'formerSmoker'])
        ]
    ))
    pop.evolve(
        initOps=[
            sim.InitSex(),
            sim.InitInfo(range(75), infoFields='age')] +
            [sim.InitGenotype(freq=[1-f, f], loci=i) for i,f in enumerate(alleleFreq)] + [
            sim.PyOperator(func=diseaseModel.initialize),
        ],
        preOps=[
            sim.InfoExec('age += 1'),
            # die of lung cancer or natural death
            sim.DiscardIf('age > age_death')
        ],
        matingScheme=sim.HeteroMating([
            sim.CloneMating(weight=-1),
            sim.RandomMating(ops = [
                sim.MendelianGenoTransmitter(),
                sim.PyOperator(func=diseaseModel.initialize)],
                subPops=[(0, 'adult')])
            ],
            subPopSize=lambda pop: pop.popSize() + popSize/75),
        postOps = [
            # update individual, currently ding nothing.
            sim.PyOperator(func=diseaseModel.updateStatus),
            # determine if someone has LC at his or her age
            sim.InfoExec('LC = age >= age_LC'),
            # get statistics about COPD and LC prevalence
            sim.Stat(pop, meanOfInfo='LC', subPops=[(0, sim.ALL_AVAIL)],
                vars=['meanOfInfo', 'meanOfInfo_sp']),
            sim.PyEval(r"'Year %d: Overall %.2f%% M: %.2f%% F: %.2f%% "
                r"NS: %.1f%%, S: %.2f%%\n' % (gen, meanOfInfo['LC']*100, "
                r"subPop[(0,3)]['meanOfInfo']['LC']*100,"
                r"subPop[(0,4)]['meanOfInfo']['LC']*100,"
                r"subPop[(0,5)]['meanOfInfo']['LC']*100,"
                r"subPop[(0,6)]['meanOfInfo']['LC']*100)"),
        ],
```

```
        gen = 100
    )


if __name__ == '__main__':
    LC_evolve(10000, [0.5, 0.1, 0.2, 0.3], LC_model(
        LC_beta0_male=0.0025, LC_beta0_female=0.0015, LC_a0=0.012,
        COPD_beta0=0.00015, COPD_a0=0.01, G1_smoking_rate=[0.3, 0.4, 0.5],
        rr_G2=[1, 1.5, 1.8], rr_G3=[1, 1.1, 1.3], rr_G4=[1, 1.5, 2],
        rr_random=1.3, rr_LC_maleSmoker=10, rr_LC_femaleSmoker=8,
        rr_COPD_smoker=6, rr_LC_COPD=2))
```

REFERENCES

1. D. B. Allison, M. C. Neale, M. I. Kezis, V. C. Alfonso, S. Heshka, and S. B. Heymsfield, Assortative mating for relative weight: genetic implications. *Behav Genet*, 26(2):103–111, 1996.
2. M. A. Speers, S. V. Kasl, D. H. Freeman, and A. M. Ostfeld, Blood pressure concordance between spouses. *Am J Epidemiol*, 123(5):818–829, 1986.
3. D. T. Redden and D. B. Allison, The effect of assortative mating upon genetic association studies: spurious associations and population substructure in the absence of admixture. *Behav Genet*, 36(5):678–686, 2006.
4. B. Amos, C. Schlötterer, and D. Tautz, Social structure of pilot whales revealed by analytical DNA profiling. *Science*, 260(5108):670–672, 1993.
5. M. Kimura and G. H. Weiss, The stepping stone model of population structure and the decrease of genetic correlation with distance. *Genetics*, 49(4):561–576, 1964.
6. W. C. Knowler, R. C. Williams, D. J. Pettitt, and A. G. Steinberg, Gm3;5,13,14 and type 2 diabetes mellitus: an association in American Indians with genetic admixture. *Am J Hum Genet*, 43(4):520–526, 1988.
7. J. K. Pritchard and P. Donnelly, Case–control studies of association in structured or admixed populations. *Theor Popul Biol*, 60(3):227–237, 2001.
8. B. Devlin and K. Roeder, Genomic control for association studies. *Biometrics*, 55(4):997–1004, 1999.
9. A. L. Price, N. J. Patterson, R. M. Plenge, M. E. Weinblatt, N. A. Shadick, and D. Reich, Principal components analysis corrects for stratification in genome-wide association studies. *Nat Genet*, 38(8):904–909, 2006.
10. X. Zhu, A. Luke, R. S. C., T. Quertermous, C. Hanis, T. Mosley, C. C. Gu, H. Tang, D. C. Rao, N. Risch, and A. Weder, Admixture mapping for hypertension loci with genome-scan markers. *Nat Genet*, 37(2):177–181, 2005.
11. M. W. Smith and S. J. O'Brien, Mapping by admixture linkage disequilibrium: advances, limitations and guidelines. *Nat Rev Genet*, 6(8):623–632, 2005.
12. D. Reich and N. Patterson, Will admixture mapping work to find disease genes? *Philos Trans R Soc Lond B Biol Sci*, 360(1460):1605–1607, 2005.

13. C. L. Pfaff, E. J. Parra, C. Bonilla, K. Hiester, P. M. McKeigue, M. I. Kamboh, R. G. Hutchinson, R. E. Ferrell, E. Boerwinkle, and M. D. Shriver, Population structure in admixed populations: effect of admixture dynamics on the pattern of linkage disequilibrium. *Am J Hum Genet*, 68(1):198–207, 2001.
14. B. S. Weir and C. C. Cockerham, Estimating F-statistics for the analysis of population structure. *Evolution*, 38(6):1358–1370, 1984.
15. J. C. Long, The genetic structure of admixed populations. *Genetics*, 127(2):417–428, 1991.
16. B. Peng and C. I. Amos, Forward-time simulations of non-random mating populations using simuPOP. *Bioinformatics*, 24(11):1408–1409, 2008.
17. H. Tang, J. Peng, P. Wang, and N. J. Risch, Estimation of individual admixture: analytical and study design considerations. *Genet Epidemiol*, 28(4):289–301, 2005.
18. J. K. Pritchard, M. Stephens, N. A. Rosenberg, and P. Donnelly, Association mapping in structured populations. *Am J Hum Genet*, 67(1):170–181, 2000.
19. M. J. Thun, S. J. Henley, and E. E. Calle, Tobacco use and cancer: an epidemiologic perspective for geneticists. *Oncogene*, 21(48):7307–7325, 2002.

APPENDIX

FORWARD-TIME SIMULATIONS USING simuPOP

A.1 INTRODUCTION

A.1.1 What is simuPOP?

simuPOP is a *general-purpose individual-based forward-time population genetics simulation environment* based on Python, a dynamic *object-oriented* programming language that has been widely used in biological studies. More specifically,

- simuPOP is a *population genetics simulator* that simulates the evolution of populations. It uses a discrete generation model, although overlapping generations could be simulated using nonrandom mating schemes.
- simuPOP explicitly models *populations with individuals* who have their own genotype, sex, and auxiliary information such as age. The evolution of a population is modeled by populating an offspring population with offspring produced from parents in this population.

Forward-time Population Genetics Simulations: Methods, Implementation, and Applications,
Bo Peng, Marek Kimmel, and Christopher I. Amos.

- Unlike coalescent-based programs, simuPOP evolves populations *forward in time*, subject to an arbitrary number of genetic and environmental forces such as mutation, recombination, migration, and population size changes.
- simuPOP is a *general-purpose* simulator that does not aim at any particular application area. It is a development tool with which a large number of simulations can be implemented. Owing to an *object-oriented design*, all classes can be extended by users to define customized genetic effects in Python. In contrast, other programs either do not allow customized effects or force users to modify code at a lower (e.g., C/C++) level.

simuPOP consists of a number of Python modules that provide a large number of Python classes and functions, including population, mating schemes, operators, (objects that manipulate populations) and simulators to coordinate the evolutionary processes. More than 70 operators are provided, covering all important aspects of genetic studies such as mutation, migration, recombination, gene conversion, natural selection, penetrance, quantitative trait, statistics calculation, and sample generation. In addition, because simuPOP provides a large number of functions to manipulate populations, it can also be used as a data manipulation and analysis tool.

Although it is generally easy to translate an evolutionary process into a simuPOP script, it can be a daunting task if the script involves complex demographic and genetic features and functions to interoperate with other applications and file formats. Fortunately, an increasing list of simuPOP functions and scripts, many of which are contributed by simuPOP users, are provided in the simuPOP online cookbook (`http://simupop.sourceforge.net/cookbook`). These recipes include functions to manipulate genetic data in other formats (e.g., the HapMap data set) [1], user-defined operators to extend the functionality of simuPOP, examples to use some of the advanced features of simuPOP, scripts to demonstrate classic population genetic models, and complete scripts that simulate a variety of evolutionary processes. It is strongly recommended that users of simuPOP make use of this resource and contribute to it whenever possible.

A.1.2 An Overview of simuPOP Concepts

A simuPOP *population* consists of *individuals* of the same *genotype structure*, which consists of properties such as number of homologous sets of chromosomes (ploidy), number of chromosomes, names and locations of markers on each chromosome, and names of auxiliary information attached to each individual (*information fields*). Individuals can be divided into *subpopulations* that can be further grouped into *virtual subpopulations* (VSPs) according to individual properties such as sex and affection status. Each population has a dictionary, called its *local namespace*, that is used to store arbitrary Python variables.

simuPOP uses *a discrete-generation model* in which the evolution of a population for one generation is characterized by the generation of an offspring population from a parental population (Figure A.1). During this process, arbitrary numbers of *operators* (Python objects that act on a population) can be applied to the parental population (*premating operator*), offspring population (*postmating operators*), or to offspring when he or she is produced (*during-mating operators*). At the end of a generation, the offspring population becomes the parental population of the next generation. This process can repeat for specified generations or be terminated if a pre- or postoperator fails to apply.

A simuPOP *mating scheme* is responsible for choosing parent or parents from a parental (virtual) subpopulation and for populating an offspring population. simuPOP provides a number of predefined *homogeneous mating schemes*, such as random, monogamous, or polygamous mating, selfing, and haplodiploid mating in hymenoptera. More complicated nonrandom mating schemes such as mating in age-structured populations can be constructed using *heterogeneous mating schemes*, which apply multiple homogeneous mating schemes to different (virtual) subpopulations.

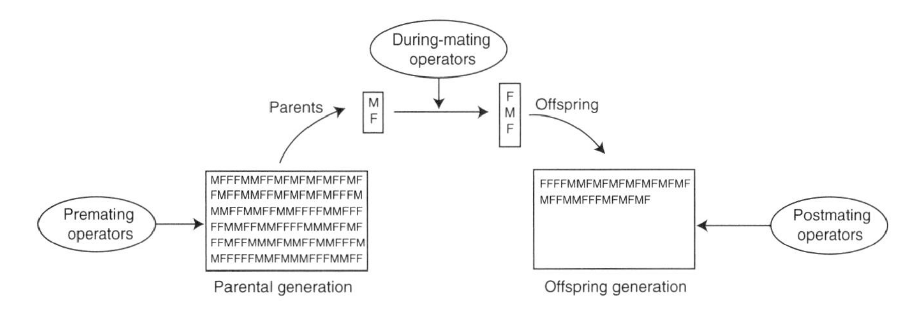

FIGURE A.1 A life cycle of an evolutionary process. Illustration of the discrete-generation evolutionary model used by simuPOP. A life cycle of a generation starts from a parental population and ends at an offspring population. Operators can be applied to the parental population before mating and to the offspring population after mating. A mating scheme is responsible for choosing parents and produce offspring. During-mating operators are used to transmit genotype and other information from parents to offspring.

SOURCE CODE A.1 A Simple Example

```
>>> import simuPOP as sim
>>> pop = sim.Population(size=1000, loci=2)
>>> pop.evolve(
...     initOps=[
...         # Initialize individuals with random sex (MALE or FEMALE)
...         sim.InitSex(),
...         # Initialize individuals with two haplotypes.
...         sim.InitGenotype(haplotypes=[[1, 2], [2, 1]])
...     ],
...     # Random mating using a recombination operator
...     matingScheme=sim.RandomMating(ops=sim.Recombinator(rates=0.01)),
...     postOps=[
...         # Calculate Linkage disequilibrium between the two loci
...         sim.Stat(LD=[0, 1], step=10),
...         # Print calculated LD values
...         sim.PyEval(r"'%2d: %.2f\n' % (gen, LD[0][1])", step=10),
...     ],
...     gen=100
... )
 0: 0.25
10: 0.22
20: 0.19
30: 0.17
40: 0.16
50: 0.17
60: 0.16
70: 0.14
80: 0.12
90: 0.10
100
```

These concepts are demonstrated in Source code A.1, where a standard diploid Wright–Fisher model with recombination is simulated. This source code records a Python interactive session where Python commands are executed immediately after they are entered. This is extremely useful for debugging and testing and for the demonstration of simuPOP features. On the other hand, you can put all the commands in a file (usually with a `.py` file extension) and execute the file in batch mode. Refer to the Python documentation for how to write a Python script.

The first line of Source code A.1 imports the standard simuPOP module. It imports simuPOP as module `sim` to make the script easier to read (e.g., `sim.InitSex` instead of `simuPOP.InitSex`). The second line creates a diploid population of 1000 individuals, each having 1 chromosome with 2 loci. The `evolve()` function evolves the population using a random mating scheme and five operators.

Operators `InitSex` and `InitGenotype` are applied at the beginning of the evolutionary process. Operator `InitSex` initializes individual sex randomly and `InitGenotype` initializes all individuals with two haplotypes `-1-2-` and `-2-1-` at equal probabilities (default value of parameter `freq`). The populations are then evolved for 100 generations. A random mating scheme is used to generate offspring by selecting male and female individuals randomly from the parental population. Instead of using the default Mendelian genotype transmitter that transmits one of the two homologous chromosomes from parents to offspring, a `Recombinator` (during-mating operator) is used to recombine parental chromosomes with a recombination rate of `0.01` before one of the recombinants is transmitted to offspring.

Two operators are applied to the offspring generation (postmating) at every 10 generations (parameter `step`). Operator `Stat` calculates linkage disequilibrium between the first and second loci. The results of this operator are stored in the local namespace of the population. The last operator `PyEval` retrieves calculated linkage disequilibrium values from the local namespace and outputs it with a generation number and a trailing new line. The result represents the decay of linkage disequilibrium of this population at 10 generation intervals. The return value of the `evolve` function, which is the number of evolved generations, is also printed.

A.2 POPULATION

Populations are the most important objects in simuPOP because all other functions and objects are designed to examine or change population properties. This section describes how to create a population, and how to examine and modify its properties using its member functions. Because there are more than 80 member functions in the `Population` class, this section lists only some of the more frequently used ones. Refer to the simuPOP reference manual for a complete list and description of all member functions.

A.2.1 Creating a Population

A `Population` object consists of one or more generations of individuals, grouped by subpopulations, and a Python dictionary to hold arbitrary variables. It is instantiated (a process to create an object from the construction function of the corresponding class) from the `__init__` function of class `Population`. Several parameters can be used to specify the genotype and population structure of a population. Acceptable parameters and their usages are listed in Table A.1.

TABLE A.1 Parameters to Create a `Population` Object

Parameter with Default Value	Usage	Examples
`size=[]`	A list of subpopulation sizes. The length of this list determines the number of subpopulations of this population	`size=2000` `size=[1000, 2000]`
`ploidy=2`	Number of homologous sets of chromosomes	`ploidy=1`
`loci=[]`	Numbers of loci on each chromosome. The length of this parameter determines the number of chromosomes	`loci=10` `loci=[20]*5`
`chromTypes=[]`	A list that specifies the type of each chromosome, which can be `AUTOSOME`, `CHROMOSOMEX`, `CHROMOSOMEY`, or `CUSTOMIZED`	`chromTypes=[AUTOSOME]*22 + [CHROMOSOMEX, CHROMOSOMEY]`
`lociPos=[]`	Positions of all loci on all chromosomes, as a list of float numbers. Default to 1, 2, ..., and so on on each chromosome	`lociPos=range(100)`
`ancGen=0`	Number of the most recent ancestral generations to keep during evolution	`ancGen=2`
`chromNames=[]`	A list of chromosome names Default to ″ for all chromosomes	`chromNames=['ch1', 'ch2']`

(*Continued*)

TABLE A.1 *(Continued)*

Parameter with Default Value	Usage	Examples
alleleNames=[]	A list or a nested list of allele names. If a list of alleles is given, it will be used for all loci in this population. Otherwise, it should specify allele names for all loci.	alleleNames=['A', 'C', 'G', 'T'] alleleNames=[['A', 'T'],['C', 'G']]
lociNames=[]	A list of names for each locus.	lociNames=['a','b']
subPopNames=[]	A list of subpopulation names. All subpopulations will have name '' if this parameter is not specified.	subPopNames=['CEU', 'YRI']
infoFields=[]	Names of information fields (named float number) that will be attached to each individual.	infoFields='fitness' infoFields=['a', 'b']

Using these parameters, small or large populations with different structures can be created. For example,

```
pop = sim.Population(size=1000, ploidy=1, loci=2)
```

creates a haploid population of 1000 individuals, each of them having 1 chromosome with 2 loci.

```
pop = sim.Population(size=[1000]*5, loci=[10]*5, infoFields='fitness')
```

creates a diploid population with 5 subpopulations, each with 1000 individuals. These individuals have 5 chromosomes, each with 10 loci and an information field named `fitness`. Note that simuPOP uses a naming convention such that plural forms of parameter names are used (e.g., `infoFields`, `loci`, `rates`) if they accept multiple inputs, although both single and list formats of inputs are acceptable by these parameters (e.g., `loci=5`, `loci=[20, 40]`).

An independent copy of a population object can be created using its `clone()` member function. This function is very useful because assignment in Python creates only a new reference to an existing object. For example, if `pop` is a `Population` object,

```
pop1 = pop
pop2 = pop.clone()
```

create a new reference of `pop` as `pop1` and a new population object `pop2`. Modifying object `pop` will modify `pop1`, but not `pop2`.

A.2.2 Genotype Structure of a Population

The genotype structure of a population includes the number of homologous copies of chromosomes, the chromosome types and names, the number of loci on each chromosome, the position and name of each locus, and the information fields attached to each individual. A number of member functions are provided to retrieve such information from a `Population` object, along with some utility functions to, for example, look up the index of a locus by its name (e.g., `locusByName()`). Table A.2 lists some of the frequently used functions.

Source code A.2 demonstrates how to create a population and use its member functions to access its structural and genotypic information. This Source code creates a diploid population with two chromosomes. Loci names are specified so that correct loci can be identified by their names even if some other loci are inserted or removed during evolution (and lead to change of loci indices). It is worth noting that following the convention of the Python programming language, *all indices in simuPOP start from 0* and *end points are not part of ranges*. That is to say, loci are indexed as `0`, `1`, `2`, ..., `n-1` if there are `n` loci, and `range(chromBegin(1), chromEnd(1))` can be used to iterate through loci on chromosome 1 (the second chromosome) because `chromEnd(1)` returns the index of the last locus on chromosome 1 plus 1.

SOURCE CODE A.2 Access Genotype Structure of a Population

```
>>> import simuPOP as sim
>>> pop = sim.Population(size=[20, 30], loci=[10, 20], lociPos=list(range(20, 50)),
...     lociNames=['loc1_%d' % x for x in range(1, 11)] +
...         ['loc2_%d' % x for x in range(1, 21)],
...     alleleNames=['A', 'C', 'G', 'T'], infoFields='a')
>>> pop.ploidy()
```

TABLE A.2 Genotype Structure-Related Member Functions

Function	Usage
`ploidy()`	Returns the number of homologous sets of chromosomes
`numChrom()`	Returns the number of chromosomes
`numLoci(chrom)`	Returns the number of loci on a chromosome
`totNumLoci()`	Returns the total number of loci on all chromosomes
`absLocusIndex(chrom, locus)`	Returns the absolute index of the `locus` locus on chromosome `chrom`
`chromLocusPair(locus)`	Returns the index of a locus on a chromosome from its absolute index
`chromBegin(chrom)`	Returns the index of the first locus on a chromosome
`chromEnd(chrom)`	Returns the index of the last locus on a chromosome plus 1
`lociNames()`	Returns the names of all loci
`locusPos(locus)`	Returns the position of a locus
`locusName(loc)`	Returns the name of a locus
`locusByName(name)`	Looks up the index of a locus from its name
`lociByNames(names)`	Looks up indices of loci from their names
`alleleName(allele, locus=0)`	Returns the name of an allele at a locus
`infoField(idx)`	Returns the name of an information field
`infoFields()`	Returns the names of all information fields
`infoIdx(name)`	Returns the index of an information field

```
2
>>> pop.numChrom()
2
>>> pop.numLoci(1)
20
>>> pop.chromBegin(1)
10
>>> pop.chromEnd(1)
```

```
30
>>> pop.totNumLoci()
30
>>> pop.chromLocusPair(22) # chromosome number and relative index
(1, 12)
>>> pop.locusName(22)
'loc2_13'
>>> pop.lociByNames(['loc1_4', 'loc2_2'])
(3, 11)
>>> pop.alleleName(1)
'C'
>>> pop.locusPos(15)
35.0
>>> pop.infoFields()
('a',)
>>> pop.infoField(0)
'a'
```

A.2.3 Subpopulations and Virtual Subpopulations

A simuPOP population can have several subpopulations. After a population is created, you can check its structural information using functions `popSize()`, `numSubPop()`, and `subPopSize(sp)`, which return the total population size, number of subpopulations, and size of a particular subpopulation, respectively. A simuPOP subpopulation is usually anonymous but a name can be assigned to a subpopulation so that it can be identified after its index has been changed due to the merge and split of other subpopulations (see Table A.3).

TABLE A.3 Population Structure-Related Member Functions

Function	Usage
`popSize()`	Return the size of a population
`numSubPop()`	Return the number of subpopulations
`subPopIndPair(idx)`	Return the subpopulation ID and relative index of an individual, given its absolute index
`subPopSize(sp)`	Return the size of a subpopulation
`subPopName(sp)`	Return the name of a subpopulation
`subPopByName(name)`	Return the index of a subpopulation from its name
`setVirtualSplitter(splitter)`	Set a virtual splitter to define virtual subpopulations
`numVirtualSubPop()`	Return the number of virtual subpopulations

Individuals in a simuPOP subpopulation can be further grouped into virtual subpopulations according to their properties. For example, all male individuals, all unaffected individuals, all individuals with information field age greater than 20, and all individuals with genotype (0, 0) at a given locus can form VSPs. VSPs do not have to add up to the whole subpopulation, nor do they have to be nonoverlapping. Unlike subpopulations that have strict boundaries, VSPs change easily with the change of individual properties.

VSPs are defined by *splitters*, which are simply *definition* of VSPs. A splitter defines the same number of VSPs in all subpopulations, although sizes of these VSPs may vary across subpopulations due to individual differences. For example, a `SexSplitter()` defines two VSPs, the first with all male individuals and the second with all female individuals; an `InfoSplitter(field='x', values=[1, 2, 4])` defines three VSPs whose members have values 1, 2, and 4 at information field x, respectively. These VSPs have their own names (e.g., 'Male' and 'Female' for two VSPs defined by a `SexSplitter`) that describe the definition by which they are defined. Several splitters are provided in simuPOP and more complex VSPs can be defined as unions or intersections of existing VSPs.

A VSP is represented by a `(sp, vsp)` pair where `sp` and `vsp` are index or name of the subpopulation and the VSP within this subpopulation. A VSP can be used in most places where a subpopulation is needed. For example, function `subPopSize([0, 'Male'])` can be used to count the number of male individuals in a subpopulation if a `SexSplitter()` is used to define VSPs by individual sex. Source code A.3 demonstrates how to apply virtual splitters to a population, how to check VSP names and sizes, and how to apply different operations to individuals in different VSPs. This Source code uses the function form of operators `InitSex`, `InitInfo`, `InitGenotype`, and `Dumper`, which will be described in detail later. Note that parameter `subPop` accepts a single subpopulation or VSP ID, and parameter `subPops` accepts a list of subpopulation or VSP IDs. Because `subPops=[0,1]` refers to two subpopulations 0 and 1, a single VSP should be specified as `subPops=[(0,1)]`.

SOURCE CODE A.3 Define and Use Virtual Subpopulations

```
>>> import simuPOP as sim
>>> import random
>>> pop = sim.Population(size=[200, 400], loci=5, infoFields='x')
>>> sim.initSex(pop)
>>> # assign random numbers using operator InitInfo
>>> sim.initInfo(pop, lambda: random.randint(0, 3), infoFields='x')
```

```
>>> # define a virtual splitter by information field 'x'
>>> pop.setVirtualSplitter(sim.InfoSplitter(field='x', values=[0, 1, 2, 3]))
>>> pop.numVirtualSubPop()    # Number of defined VSPs
4
>>> pop.subPopName([0, 0])    # Each VSP has a name
'x = 0'
>>> pop.subPopSize([0, 0])    # Size of VSP 0 in subpopulation 0
62
>>> pop.subPopSize([1, 0])    # Size of VSP 1 in subpopulation 1
110
>>> # define a virtual splitter by sex
>>> pop.setVirtualSplitter(sim.SexSplitter())
>>> pop.numVirtualSubPop()    # Number of defined VSPs
2
>>> pop.subPopName([0, 0])    # Each VSP has a name
'Male'
>>> pop.subPopSize([0, 1])    # Size of VSP 0 in subpopulation 0
109
>>> # initialize male and females with different genotypes.
>>> sim.initGenotype(pop, genotype=[0]*5, subPops=[(0, 0)])
>>> sim.initGenotype(pop, genotype=[1]*5, subPops=[(0, 1)])
>>> sim.dump(pop, max=6, subPops=0, structure=False)
SubPopulation 0 (), 200 Individuals:
   0: FU 11111 | 11111 |  1
   1: FU 11111 | 11111 |  0
   2: MU 00000 | 00000 |  3
   3: MU 00000 | 00000 |  1
   4: MU 00000 | 00000 |  1
   5: MU 00000 | 00000 |  0
```

A.2.4 Accessing Individuals in a Population

Individuals are building blocks of a population. An individual object cannot be created independently from a population, but references to individuals can be retrieved using member functions of a population object. More specifically,

- `pop.individual(idx)` returns a reference to the idx-th individual in a population object `pop`.
- `pop.individuals()` returns an iterator that iterates through all individuals in this population (e.g., "`for ind in pop.individuals()`").
- `pop.individuals(subPop)` returns an iterator that iterates through individuals within a (virtual) subpopulation.

The individual objects returned by these functions are instances of the `Individual` class. They have access to all genotypic structure-related functions listed in Table A.2 (technically speaking, both the `Individual` and `Population` classes are derived from the `GenoStruTrait` class,

TABLE A.4 Member Functions of the `Individual` Class

Function	Usage
`sex()`	Return the name of a subpopulation
`setSex(sex)`	Return the index of a subpopulation from its name
`affected()`	Set a virtual splitter to define virtual subpopulations
`setAffected()`	
`allele(idx, ploidy=-1, chrom=-1)`	Return the size of a population
`setAllele(allele, idx, ploidy=-1, chrom=-1)`	Return the number of subpopulations
`genotype(ploidy=ALL_AVAIL, chroms=ALL_AVAIL)`	Return the subpopulation ID and relative index of an individual, given its absolute index
`setGenotype(geno, ploidy=ALL_AVAIL, chroms=ALL_AVAIL)`	Return the size of subpopulation
`info(field)`	Return the number of virtual subpopulations.
`setInfo(value, field)`	Remove specified subpopulations

so they both have access to all member functions of the base class) and to some member functions to read and write individual sex, affection status, genotype, and information fields (Table A.4). For example, function `Individual.sex()` returns the sex of an individual, which can be `MALE` or `FEMALE`.

From a user's point of view, genotypes of an individual are stored sequentially and can be accessed locus by locus or in batch. The alleles are arranged by position, chromosome, and ploidy. That is to say, the first allele on the first chromosome of the first homologous set is followed by alleles at other loci on the same chromosome, then alleles on the second and later chromosomes, followed by alleles on the second homologous set of the chromosomes for a diploid individual. A consequence of this memory layout is that alleles at the same locus of a nonhaploid individual are separated by `Individual.totNumLoci()` loci.

simuPOP provides several functions to read and write individual genotypes. For example, functions `Individual.allele()` and `Individual.setAllele()` can be used to read and write single alleles; functions `Individual.genotype()` and

`Individual.setGenotype()` can be used to read and write individual genotypes in batch mode. The `setGenotype` function accepts a list of alleles, which will be reused if its length is less than the total number of required alleles. For example, you can quickly set all alleles of an individual `ind` to 1 using function `ind.setGenotype(1)`.

Individual information fields can be accessed using functions `Individual.info(field)` and `Individual.setInfo(value, field)`, or as attributes of an `Individual` object. For example, if an individual `ind` has information field `id`, you can read its value using `ind.id` and set its value using statement `ind.id=55`. Source code A.4 demonstrates how to access and modify individual sex, affection status, and information fields using these functions.

SOURCE CODE A.4 Access to Individuals in a Population

```
>>> import simuPOP as sim
>>> import random
>>> pop = sim.Population(size=[4, 5], loci=4, infoFields='a')
>>> # iterate through all individuals in the first subpopulation of pop
>>> for ind in pop.individuals(0):
...     ind.setSex(sim.FEMALE)
...     ind.setAllele(1, idx=1, ploidy=1)
...     ind.a = random.randint(2, 5)
...
>>> ind = pop.individual(2)
>>> ind.sex() # The numeric value of FEMALE is printed
2
>>> ind.genotype()
[0, 0, 0, 0, 0, 1, 0, 0]
>>> ind.setGenotype([1, 0, 1])
>>> ind.genotype()
[1, 0, 1, 1, 0, 1, 1, 0]
>>> ind.a
5.0
>>> ind.a = 10
>>> ind.a
10.0
```

A.2.5 Population Variables

Each simuPOP population has a Python dictionary that can store arbitrary Python variables. These variables are usually used by various operators to set and retrieve population statistics. For example, the `Stat` operator calculates population statistics and stores the results in this dictionary. Other operators such as the `PyEval` and `TerminateIf` read from this dictionary and act upon its values.

The `Population` class provides two member functions, namely, `Population.vars()` and `Population.dvars()` to access a population dictionary. These functions return the same dictionary object, but `dvars()` returns a wrapper class so that you can access keys in this dictionary as attributes. For example, `pop.vars()['alleleFreq'][0]` is equivalent to `pop.dvars().alleleFreq[0]`.

It is important to understand that this dictionary forms a *local namespace* in which Python expressions can be evaluated. That is to say, *items in this dictionary can be treated as variables in a namespace and be used to execute Python statements and expressions*. This is the basis of how expression-based operators work. For example, the `PyEval` operator in Source code A.1 evaluates an Python expression

```
'%d: %.2f\n' % (gen, LD[0][1])
```

in a population's local namespace when it is applied to that population. This expression uses two variables `gen` and `LD`. Variable `gen` is created and maintained automatically during the evolution of a population. It records the current generation of the population. Variable `LD` is set by operator `Stat` when it is applied to the population before operator `PyEval` is applied. Because arbitrary Python expressions can be evaluated, a `PyEval` operator can output these statistics or their derived values in any format. For example, expression

```
'1-sum([x*x for x in alleleFreq[0].values()])'
```

calculates the expected heterozygosity ($H = 1 - \sum p_i^2$ where p_i is the allele frequency of allele i) at locus 0 after a dictionary of allele frequencies `alleleFreq` is calculated by operator `Stat(alleleFreq=0)`.

A.2.6 Altering the Structure, Genotype, or Information Fields of a Population

The `Population` class provides a number of member functions to alter the structure of a population and to access individual genotype and information fields in batch mode. Table A.5 lists some of the frequently used functions. Complete prototypes are ignored because some of them accept many parameters.

Function `removeIndividuals` removes individuals from a population according to their indices, IDs (value at an information field), or the return values of a filter function that accepts each individual as its input value.

TABLE A.5 Population Modification Functions of the `Population` Class

Function	Usage
`removeSubPops`	Remove specified subpopulations
`mergeSubPops`	Merge specified subpopulations into one subpopulation
`splitSubPop`	Split a subpopulation into subpopulations with given sizes
`resize`	Resize a population with new subpopulation sizes
`removeIndividuals`	Remove individuals by indices, IDs, or a filter function
`addIndFrom`	Add individuals with the same genotype structure from another population
`addChrom`	Add a chromosome to the current population
`addChromFrom`	Add chromosomes from another population with the same number of individuals
`addLoci`	Add some loci to the current population
`addLociFrom`	Add loci from another population with the same number of individuals
`removeLoci`	Remove selected loci from a population
`genotype`	Return a list-like object that represents genotypes of all individuals in a (sub)population
`setGenotype`	Set genotype of all individuals in a (sub)population using a lists of alleles
`addInfoFields`	Add additional infomation fields to a population
`removeInfoFields`	Remove specified information fields of a population
`indInfo`	Return values of an information field of all individuals or individuals in a (virtual) subpopulation
`setIndInfo`	Set values of an information field of all individuals or individuals in a (virtual) subpopulation using a list of values

Alternatively, function `removeSubPops` can be used to remove subpopulations or groups of individuals who share the same properties (use VSP). Functions `extractIndividuals` and `extractSubPops` work similarly. However, instead of removing selected individuals or (virtual) subpopulations, they copy these individuals and form a new population from them. These functions are usually used to draw samples from an existing population.

Functions `mergeSubPops` and `splitSubPop` merge and split existing subpopulations and are usually used to merge or split parental populations during the simulation of demographic models with population merge and split. If population size needs to be changed, the function `resize` can resize a population and the function `addIndFrom` can merge individuals from another population to the current population.

When simuPOP is used to process real empirical data sets, functions such as `removeLoci`, `addLociFrom`, and `addChromFrom` can be used to remove loci or combine data sets with different loci. We will not go into the details of these functions because they are used primarily for data processing, which is not the focus of this book. Refer to scripts such as `loadHapMap3.py` (a script to import the HapMap data sets and save them in simuPOP formats) in the simuPOP online cookbook for examples on how to use these functions.

Individuals in a population share the same set of information fields, so the addition and removal of information fields can be performed only at the population level. Because the unused information fields tend to hinder the efficiency of simulations, it is a common practice to use minimal set of information fields during evolution and add additional information fields using function `addInfoFields` for postevolution data analysis.

Finally, the `Population` class provides functions to access individual genotype and information fields in batch mode. For example, function `indInfo` returns values of an information field of all individuals or individuals belonging to a (virtual) subpopulation. This makes it easy to calculate summary statistics of these information fields. Similarly, function `genotype` returns genotypes of all or individuals in certain (virtual) subpopulation so that you can count, for example, the number of mutants in a population. For performance considerations, function `genotype` returns a special object that directly exposes the underlying genotypes to users. Modifying this object will change individual genotypes.

Source code A.5 demonstrates how to use some of the mentioned functions.

SOURCE CODE A.5 Use of Population Modification and Batch Access Functions

```
>>> import simuPOP as sim
>>> import random
>>> pop = sim.Population(size=[4, 6], loci=2, infoFields='x')
>>> pop.setIndInfo([random.randint(0, 10) for x in range(10)], 'x')
>>> sum(pop.indInfo('x')) / pop.popSize()  # get the mean of field x
6.0
>>> pop.setGenotype([0, 1, 2, 3], 0)
>>> pop.genotype(0)  # for the first subpopulation
[0, 1, 2, 3, 0, 1, 2, 3, 0, 1, 2, 3, 0, 1, 2, 3]
>>> pop.setVirtualSplitter(sim.InfoSplitter(cutoff=[3], field='x'))
>>> pop.mergeSubPops(subPops=[0,1]) # merge two subpopulations
```

```
0
>>> pop.setGenotype([0])     # clear all values
>>> pop.setGenotype([5, 6, 7], [0, 1])  # only for x >= 3
>>> pop.indInfo('x', 0)
(8.0, 9.0, 10.0, 8.0, 10.0, 5.0, 0.0, 1.0, 2.0, 7.0)
>>> pop.removeSubPops([(0,0)])          # remove individuals with x < 3
>>> pop.popSize()
7
>>> pop.genotype(0) # so all existing individuals have genotype 5, 6, 7
[5, 6, 7, 5, 6, 7, 5, 6, 7, 5, 6, 7, 5, 6, 7, 5, 6, 7, 5, 6, 7, 5, 6, 7, 5, 6, 7, 5]
```

A.2.7 Multigeneration Populations and Parental Information

All simulations we have described so far discard parental information. That is to say, a parental population is discarded when it is replaced by its offspring population at the end of an evolutionary cycle. This behavior can be changed by setting the *ancestral depth* of a population, namely, how many ancestral generations to keep during an evolutionary process. For example, a population

```
Population(10000, loci=5, ancGen=2)
```

created in Source code A.6 will keep its parental (`ancGen=1`) and grandparental (`ancGen=2`) generations during an evolutionary process. At the end of each generation, the existing grandparental generation is dicarded, the parental generation becomes the grandparental generation, the present population becomes the parental generation, and the offspring population becomes the present generation. After 20 generations, the population object will have 3 generations, namely, the offspring population at the end of generation 17, 18, and 19. These generations could be set as the present population using function `Population.useAncestralGen(gen)` where `gen` is `0` for present, `1` for parental, and `2` for grandparental generation.

SOURCE CODE A.6 Keeping Multiple Ancestral Generations During an Evolutionary Process

```
>>> import simuPOP as sim
>>> pop = sim.Population(size=10000, loci=5, ancGen=2)
>>> pop.evolve(
...     initOps=[
...         sim.InitSex(),
...         sim.InitGenotype(freq=[0.7, 0.3]),
...     ],
...     matingScheme=sim.RandomMating(),
...     postOps=[
...         sim.MaPenetrance(loci=2, penetrance=[0.05, 0.1, 0.12]),
...     ],
```

```
...     gen=20
... )
20
>>> for gen in range(pop.ancestralGens() + 1):
...     pop.useAncestralGen(gen)
...     sim.stat(pop, numOfAffected=True)
...     print('Number of affected individuals in generation %d is %d' % \
...         (gen, pop.dvars().numOfAffected))
...
Number of affected individuals in generation 0 is 760
Number of affected individuals in generation 1 is 773
Number of affected individuals in generation 2 is 787
```

Although the multigeneration population created in Source code A.6 keeps two ancestral generations, it does not keep any parental information, so it is not possible to identify parents of each individual. In order to keep parental information, it is necessary to assign unique IDs to all individuals and store IDs of parents to their offspring. simuPOP reserves information fields `ind_id`, `father_id` and `mother_id` and operators `IdTagger` and `PedigreeTagger` for such purposes. More specifically, in order to track parentship of a population, you should add information fields `ind_id`, `father_id`, and `mother_id` to the population and use operator `IdTagger` to assign a unique ID to field `ind_id` of each individual and an operator `PedigreeTagger` to record the ID of the father and mother of each offspring to fields `father_id`, and `mother_id` respectively.

Source code A.7 demonstrates how to assign IDs and record parentship of each individual during evolution. In this Source code, an operator `IdTagger` is used to initialize all individuals in the starting population with IDs 1, 2, ..., 1000. During evolution, operator `IdTagger` assigns a unique ID to each offspring and operator `PedigreeTagger` copies `ind_id` of his or her parents to fields `father_id` and `mother_id`. If we examine the value of these information fields of the simulated population, we can see that individuals have unique IDs, and two individuals share the same parents because each mating event produces two offspring (`numOffspring=2`).

SOURCE CODE A.7 Recording Parentship of Individuals During Evolution

```
>>> import simuPOP as sim
>>> pop = sim.Population(size=1000, loci=5, ancGen=2,
...     infoFields=['ind_id', 'father_id', 'mother_id'])
>>> pop.evolve(
...     initOps=[
...         sim.InitSex(),
...         sim.InitGenotype(freq=[0.7, 0.3]),
...         sim.IdTagger(),
```

```
...     ],
...     matingScheme=sim.RandomMating(ops=[
...         sim.MendelianGenoTransmitter(),
...         sim.IdTagger(),
...         sim.PedigreeTagger()],
...         numOffspring=2),
...     postOps=[
...         sim.MaPenetrance(loci=2, penetrance=[0.05, 0.1, 0.12]),
...     ],
...     gen=20
... )
20
>>> pop.indInfo('ind_id')[:5]
(20001.0, 20002.0, 20003.0, 20004.0, 20005.0)
>>> pop.indInfo('father_id')[:5]
(19443.0, 19443.0, 19838.0, 19838.0, 19755.0)
>>> pop.indInfo('mother_id')[:5]
(19181.0, 19181.0, 19854.0, 19854.0, 19756.0)
```

A.2.8 Saving and Loading a Population

A population can be saved to a disk file using function `Population.save(filename)`, and be loaded from this file using global function `loadPopulation(filename)`. simuPOP uses a binary format to save a population object. The format is not human readable, but is portable in the sense that a file saved in one platform can be loaded by simuPOP on another platform. The simuPOP cookbook provides a number of functions to save and load simuPOP populations in formats used by other genetic analysis software.

A.3 OPERATORS

Operators are objects that act on populations. During an evolutionary process, operators are applied to populations repeatedly, just like what robots do in an automotive production line. There are two types of operators: operators that are applied to populations before or after mating, and operators that are applied to offspring during mating. Some operators could be applied to both populations and individuals to perform different tasks.

Operators that are applied to populations are used in parameters `initOps`, `preOps`, `postOps`, and `finalOps` of the `Population.evolve()` function. The `initOps` operators are applied before an evolutionary process, the `preOps` operators are applied to the parental population at each generation before mating, the `postOps` operators are applied to the offspring population at each generation after mating, and the `finalOps` operators are applied after an evolutionary process. These operators include fitness operators that set individual fitness values before mating, mutation operators that mutate alleles, statistics calculators that calculate population statistics, and operators that report the progress of an evolutionary process.

Operators that are applied to individuals are used in the `ops` parameter of a mating scheme. They are usually used to transmit genotype or other information from parents to offspring. Examples of such operators include `MendelianGenoTransmitter` that transmits parental genotype to offspring according to Mendelian laws and `PedigreeTagger` that records the IDs of parents to each offspring.

A.3.1 Applicable Generations

Operators are, by default, applied to all generations during an evolutionary process. This can be changed using the `begin`, `end`, `step`, and `at` parameters of operators. As their names indicate, these parameters control the starting generation (`begin`), ending generation (`end`), generations between two applicable generations (`step`), and an explicit list of applicable generations (`at`, a single generation number is also acceptable). Other parameters will be ignored if parameter `at` is specified. If the number of generations to evolve is fixed (parameter `gen` of the `Population.evolve` function is specified), negative generation numbers are allowed. They are counted backward from the ending generation. For example, if a simulation starts at generation `0`, and the `evolve` function has parameter `gen=10`, the simulator will stop at the *beginning* of generation `10`. Generation `-1` refers to generation `9` (the last generation), and generation `-2` refers to generation `8`, and so on.

These parameters give simuPOP the flexibility to apply operators at selected generations. For example, you can calculate and output statistics at every 10 generations to avoid excessive report or apply different migration models during different stages of an evolutionary process. Source code A.8 demonstrates how to set applicable generations of an operator. In this Source code, operators `InitSex` ad `InitGenotype` are applied once before evolution. Operators `Stat` and `PyEval` are applied at every 10 generations, namely, generations 0, 10, 20, ... At these generations, operator `Stat` is applied before mating (on the parental population) to calculate allele frequency at locus 0, and `PyEval` is applied after mating (on the offspring population) to output variable `alleleFreq[0][1]` where `0` and `1` are locus and allele indices respectively. Although `PyEval` is applied after mating, the allele frequencies it reports are actually frequencies of the parental population calculated by operator `Stat` before mating.

An operator `MapSelector` is used to select against allele 1 starting from generation 50, so the frequency of allele 1 decreases quickly after that generation.

SOURCE CODE A.8 Applicable Generations of an Operator

```
>>> import simuPOP as sim
>>> pop = sim.Population(size=2000, loci=1, infoFields='fitness')
>>> pop.evolve(
...     initOps=[
...         sim.InitSex(),
...         sim.InitGenotype(freq=[0.5, 0.5]),
...     ],
...     preOps=[
...         sim.Stat(alleleFreq=0, step=10),
...         sim.MapSelector(begin=50, loci=0,
...             fitness={(0,0):1, (0,1):0.99, (1,1):0.97})
...     ],
...     matingScheme=sim.RandomMating(),
...     postOps=sim.PyEval(r"'%3d: %.2f\n' % (gen, alleleFreq[0][1])",
...         step=10),
...     gen=100
... )
  0: 0.50
 10: 0.52
 20: 0.52
 30: 0.50
 40: 0.48
 50: 0.48
 60: 0.46
 70: 0.46
 80: 0.44
 90: 0.37
100
>>>
```

A.3.2 Operator Output

All operators we have seen write their output to the standard output, namely, a terminal window. However, in a complex evolutionary system where multiple statistics are recorded, you might want to store certain statistics in one file and others in another file. In these cases, you can use parameter `output` of operators to direct their output to other destinations.

Parameter `output` accepts an output specification string or a user-defined Python function (Table A.6). Because statistics are usually collected and outputted repeatedly during evolution, the most frequently used format is `'>>filename'`. Files specified in this way will be opened before evolution, accept inputs from one or more operators during evolution, and be closed afterward. A comma or tab separated file can be created in this way if proper delimiters are outputted.

TABLE A.6 Acceptable Inputs for Parameter `output` of Operators

Parameter	Usage
''	Supress output
'filename'	Write output to a file named filename and close it Existing content of this file will be cleared
'>filename'	Equivalent to 'filename'
'>>filename'	Append output to a file named filename. Clear the file before evolution if this file already exists
'>>>filename'	Append output to a file named filename. Do not clear the file before evolution if this file already exists
'!expr'	Obtain an output specification string by evaluating expression expr in the local namespace of the current population
a Python function	Send the output to a user-defined Python function

A output specification will be considered as an expression if it starts from an exclamation symbol. Such an expression will be evaluated in a population's local namespace to determine a proper output. For example, parameter `output='!"gen_%d.txt" % gen'` directs output from an operator to files `gen_0.txt`, `gen_1.txt` etc at generations 0, 1, ... As an advanced feature, operator output could be sent directly to a Python function for real time analysis.

Source code A.9 demonstrates how to use the `output` parameters to record two LD measures to files `LD.txt` and `R2.txt` separately.

SOURCE CODE A.9 Use of Parameter Output of Operators to Redirect Operator Output

```
>>> import simuPOP as sim
>>> pop = sim.Population(size=2000, loci=2)
>>> pop.evolve(
...     initOps=[
...         sim.InitSex(),
...         sim.InitGenotype(genotype=[1, 2, 2, 1])
...     ],
...     matingScheme=sim.RandomMating(ops=sim.Recombinator(rates=0.1)),
...     postOps=[
...         sim.Stat(LD=[0, 1]),
...         sim.PyEval(r"'%3d: %.4f\n' % (gen, LD[0][1])",
...             output='>>LD.txt'),
...         sim.PyEval(r"'%3d: %.4f\n' % (gen, R2[0][1])",
...             output='>>R2.txt')
```

```
...     ],
...     gen=100
... )
100
>>> # print the first five lines of the output
>>> print(''.join(open('R2.txt').readlines()[:5]))
  0: 0.6123
  1: 0.5055
  2: 0.4075
  3: 0.3513
  4: 0.2808
```

A.3.3 During-Mating Operators

Source code A.1 uses operator `Recombinator(rate=0.01)` in the `ops` parameter of mating scheme `RandomMating`. This operator retrieves parental chromosomes, recombines them with specified recombination rate, and passes one recombinant from each parent to the offspring. The random mating scheme in Source code A.8 does not specify an `ops` parameter so that a default during- mating operator for this mating scheme, namely, a `MendelianGenoTransmitter()`, is used to transmit genotype from parents to offspring according to Mendelian laws. These during-mating operators are called *genotype transmitters* just to indicate they are responsible for transmitting parental genotypes to offspring.

In addition to genotype transmitters, other during-mating operators could be applied during the production of offspring. For example, operator `PedigreeTagger` records the IDs of parents to information fields of offspring, and operator `InheritTagger` passes parental information fields to offspring. It is important to remember that a genotype transmitter needs to be explicitly specified when the `ops` parameter is used.

Source code A.10 uses an operator `InheritTagger` to record the ancestry of each individual. The simulation starts from a population of 1000 individuals, half with ancestry value 0 and half with ancestry value 1. During evolution, a `MendeliangenoTransmitter` passes parental genotypes and a `InheritTagger` passes the mean of ancestral ancestry values to their offspring. Because a random mating scheme selects parents randomly, it is not surprising that the majority of the individuals have an ancestry value around 0.5 after only a few generations. This Source code uses operator `InitInfo` to assign a sequence of values (`[0,1]`) to individuals in a population, and a population member function `Population.indInfo` to get the values of information field `anc` of all individuals.

SOURCE CODE A.10 Use of an InheritTagger to Track Individual Ancestry

```
>>> import simuPOP as sim
>>> pop = sim.Population(size=1000, loci=20, infoFields='anc')
>>> pop.evolve(
...     initOps=[
...         sim.InitSex(),
...         sim.InitGenotype(freq=[0.5, 0.5]),
...         sim.InitInfo([0,1], infoFields='anc'),
...     ],
...     matingScheme=sim.RandomMating(ops=[
...         sim.MendelianGenoTransmitter(),
...         sim.InheritTagger(mode=sim.MEAN, infoFields='anc')]),
...     gen=10
... )
10
>>> # find the ancestral values
>>> anc = pop.indInfo('anc')
>>> print('min anc: %.3f, mac anc: %.3f' % (min(anc), max(anc)))
min anc: 0.432, mac anc: 0.558
```

A.3.4 Function Form of Operators

Operators are usually applied to populations during evolution, although they can also be applied to a population directly, using their function counterparts. These functions are named similar to the corresponding classes. They take a population object as their first parameter, create an operator using the rest of the parameters, and apply the operator to the passed population. For example, operators used in the `initOps` parameter of Source code A.10 can be moved before function `pop.evolve` to initialize the population as follows:

```
pop = sim.Population(size=1000, loci=20, infoFields='anc')
sim.initSex(pop)
sim.initGenotype(pop, freq=[0.5, 0.5])
sim.initInfo(pop, [0, 1], infoFields='anc')
pop.evolve( ... )   # ignored
```

Function `stat` is the function form of operator `Stat`, which is frequently used to calculate statistics of populations. For example, it is more efficient to calculate the minimum and maximum of ancestry values of the simulated population in Source code A.10 using

```
stat(pop, minOfInfo='anc', maxOfInfo='anc')
print('min anc: %.3f, max anc: %.3f' % \
    (pop.dvars().minOfInfo['anc'], (pop.dvars().minOfInfo['anc']))
```

because the `Stat` operator calculates the statistics internally without copying values of information field `'anc'` to a list.

A.3.5 Operator `Stat`

Operator `Stat` is used to calculate statistics for populations, subpopulations, or groups of individuals in subpopulations that share certain properties (virtual subpopulations). Instead of returning calculated statistics in some way, this operator stores one or more variables in a population's local namespace. These variables can be accessed by other operators such as `PyEval` or directly from a population object using member function `Population.vars()` or `Population.dvars()`.

Table A.7 lists acceptable parameters of the `Stat` operator and the variables it sets for each statistics. Variables marked by an asterisk ($*$) are the default variables that will be set for a statistic. An alternative set of variables can be specified via parameter `vars` of this operator. For example, operator

```
Stat(association=ALL_AVAIL, vars='Armitage_p')
```

performs Cochran–Armitage trend tests at all available loci of a population and sets a dictionary `Armitage_p` for the *p*-values of the tests. The default allele-based association tests will not be performed in this case.

This operator by default does not calculate statistics for subpopulation. If you list a set of (virtual) subpopulations using parameter `subPops`, individuals from these subpopulations will be pulled together for calculation. For example,

```
pop.setVirtualSplitter(SexSplitter())
stat(pop, alleleFreq=0, subPops=[(ALL_AVAIL, 'Male')])
```

calculates allele frequency at locus 0 for all male individuals in all subpopulations of population `pop`. The results will be saved to variables `allele-Freq` and `alleleNum`, although they are the allele frequency among all male individuals instead of the allele frequency for all individuals in a population. This example uses `[(ALL_AVAIL, 'Male')]` to represent the first virtual subpopulations in all available subpopulations. Similarly, `[(0, ALL_AVAIL)]` or `[(ALL_AVAIL, ALL_AVAIL)]` can be used to refer to all virtual subpopulations in a specific subpopulation or in all subpopulations.

Subpopulation-specific statistics can be calculated for variables marked by a *s* symbol in Table A.7, by specifying variables with a `_sp` suffix in parameter `vars` of the `Stat` operator. The resulting variables will be stored in dictionaries `subPop[sp]` where `sp` is the ID of (virtual) subpopulations. For example, operator

TABLE A.7 Parameters and Variables of Operator `Stat`

Statistics	Parameter	Variables
Population size	`popSize=False`	`popSize`[*si], `subPopSize`[l]
Number of male individuals	`numOfMales=False`	`numOfMales`[*si], `numOfFemales`[*si], `propOfMales`[sf], `propOfFemales`[sf]
Number of affected individuals	`numOfAffected=False`	`numOfAffected`[*si], `numOfUnaffected`[*si], `propOfAffected`[sf], `propOfUnaffected`[sf]
Allele count and frequency	`alleleFreq=[]`	`alleleFreq`[*sd], `alleleNum`[*sd]
Heterzygote frequency	`heteroFreq=[]`	`heteroFreq`[*sd], `heteroNum`[sd]
Homozygote frequency	`homeFreq=[]`	`homoFreq`[*sd], `homoNum`[sd]
Genotype frequency	`genoFreq=[]`	`genoFreq`[*sd], `genoNum`[*sd]
Haplotype frequency	`haploFreq=[]`	`haploFreq`[*sd], `haploNum`[*sd]
Sum of information fields	`sumOfInfo=[]`	`sumOfInfo`[*sd]
Mean of information fields	`meanOfInfo=[]`	`meanOfInfo`[*sd]
Variance of information fields	`varOfInfo=[]`	`varOfInfo`[*sd]
Maximum of information fields	`maxOfInfo=[]`	`maxOfInfo`[*sd]
Minimum of information fields	`minOfInfo=[]`	`minOfInfo`[*sd]
Linkage disequilibrium	`LD=[]`	`LD`[*sd], `LD_prime`[*sd], `R2`[*sd], `LD_ChiSq`[sd], `LD_ChiSq_p`[sd], `CramerV`[sd]
Association tests	`association=[]`	`Allele_ChiSq_p`[*sd], `Allele_ChiSq`[sd], `Geno_ChiSq`[sd], `Geno_ChiSq_p`[sd], `Armitage_p`[sd]
Neutrality tests	`neutrality=[]`	`Pi`[*sf]
Population structure	`structure=[]`	`F_st`[*f], `F_is`[f], `F_it`[f], `f_st`[d], `f_is`[d], `f_it`[d], `G_st`[f], `g_st`[d]
Hardy–Weinberg equilibrium	`HWE=[]`	`HWE`[*sd]

*: Default variable, *s*: available for (virtual) subpopulations if `varname_sp` is specified. *i*: integer variable, *f*: float variable, *d*: dictionary variable.

```
Stat(alleleFreq=0, vars='alleleFreq_sp')
```

calculates allele frequencies at locus 0 for all subpopulations and sets variables such as `subPop[0]['alleleFreq']`, whereas

```
Stat(alleleFreq=0, subPops=[(ALL_AVAIL, 0)], vars='alleleFreq_sp')
```

calculates allele frequencies for specified virtual subpopulations and sets variables such as `subPop[(1,0)]['alleleFreq']`. Note that `pop.dvars(sp).alleleFreq` can be used as a shortcut to access variable `pop.dvars().subPop[sp]['alleleFreq']`.

Variables set by the Stat operator can be an integer (marked by a *i* symbol in Table A.7), a float number (*f*), a list of numbers (*l*), and a dictionary (*d*). The keys of these dictionaries vary from statistic to statistic. For example, variable `meanOfInfo['fitness']` records the mean of information fitness; variable `alleleFreq[0][a]` records the frequency of allele `a` at locus `0` using alleles as keys; and variable `haploNum[(0,1,2)][(0,0,1)]` records the frequency of haplotype `-0-0-1-` at loci `0`, `1`, and `2`, using a tuple of alleles as keys. Because keys of these dictionaries sometimes cannot be determined in advance, access to these dictionaries with an invalid key will return 0 instead of triggering an `KeyError` exception. For example, if there are alleles 0, 2, 3 at locus 1, operator

```
Stat(alleleFreq=1)
```

will set dictionaries `alleleFreq[1]` and `alleleNum[1]` with keys `0`, `2`, and `3`. You can use expression `len(alleleFreq[1])` to obtain the number of alleles at this locus, expression `alleleFreq[1].keys()` to obtain a list of available alleles, and most importantly, expressions such as

```
alleleFreq[1][1]
```

to output or display frequency of a particular allele without worrying about whether or not there is such an allele at this locus.

A.3.6 Hybrid and Python Operators

Given the large number of population genetics models and statistics, it is not possible for simuPOP to provide native support for all of them. Fortunately, owing to the scripting language design, it is easy to extend functions of simuPOP in Python through the use of user-provided *callback functions*.

A simuPOP *callback function* is a user-defined Python function that is passed to and called by simuPOP. The interface of this function is specified by simuPOP. For example, a demographic function is a callback function that is passed to the `subPopSize` parameter of a mating scheme and is called by the mating scheme at each generation before mating happens. This function accepts a generation number with parameter `gen` and/or a parental population with parameter `pop` and returns the population size (if there is no population structure) or a list of subpopulation sizes of the offspring population. It is noted that *simuPOP depends on parameter names to determine what should be passed to a callback function*. For example, a mating scheme will pass a parental population to function `demo(pop)`, a generation number to function `demo(gen)`, and both parental population and generation number to function `demo(pop, gen)`.

A *hybrid operator* is an operator that accepts a callback function (Does it accept something else, since it is called a hybrid?). The number and meaning of input parameters and return values vary from operator to operator. For example, a hybrid mutator sends a to-be-mutated allele to a callback function and uses its return value as the mutant allele. A hybrid selector uses return values of a user-defined function as individual fitness values. Such an operator handles the routine part of the work (e.g., scan through a chromosome and determine which allele needs to be mutated) and leaves the creative part to users. For example, Source code A.11 defines an asymmetric stepwise mutation model with random steps using a hybrid mutator called `PyMutator`. This mutator mutates alleles at loci 2 and 5 with specified mutation rates and sends alleles to be mutated to a callback function `randomStep`, which mutates an allele a to allele $a-1$, a, $a+1$, $a+2$ with equal probabilities. Because of the upward trend of this mutational process, the average number of tandem (why are they tandem repeats, this seems to be single repeat differences between alleles) repeats increases during evolution.

SOURCE CODE A.11 An Asymmetric Stepwise Mutation Model with Random Steps

```
>>> import simuPOP as sim
>>> import random
>>> def randomStep(allele):
...     return allele + random.randint(-1, 2)
...
>>> pop = sim.Population(size=1000, loci=[10])
```

```
>>> pop.evolve(
...     initOps=[
...         sim.InitSex(),
...         sim.InitGenotype(genotype=100)
...     ],
...     matingScheme=sim.RandomMating(),
...     postOps=sim.PyMutator(func=randomStep, rates=[1e-3, 1e-2],
...             loci=[2, 5]),
...     gen = 1000
... )
1000
>>> # count the average number of tandem repeats at both loci
>>> sim.stat(pop, alleleFreq=[2, 5])
>>> sum([x*y for x,y in pop.dvars().alleleNum[2].items()])/(2.*pop.popSize())
99.859
>>> sum([x*y for x,y in list(pop.dvars().alleleNum[5].items())])/(2.*pop.popSize())
104.9455
```

A Python operator `PyOperator` is the most flexible hybrid operator in simuPOP because its callback function takes a population or an individual as its input and can perform arbitrary operations on them. When this operator is applied to a parental or offspring population, it passes the population directly to a callback function with an optional parameter (parameters `pop` and `param`). For example, operator `PyOperator` used in Source code A.12 passes offspring populations to a function `drawSample` at every 100 generations. This function draws cases and controls from these populations and saves samples for later analysis. Numbers of cases and controls are passed to this function using parameter `param`. Note that a callback function for operator `PyOperator` must return `True` or `False` and the evolution of a population will be terminated if `False` is returned.

SOURCE CODE A.12 Use of a Python Operator to Draw Sample at Every 100 Generations

```
import simuOpt
simuOpt.setOptions(quiet=True, alleleType='binary')
import simuPOP as sim
from simuPOP.sampling import drawCaseControlSample

def drawSample(pop, param):
    'Rest fixed locus to all zero alleles'
    nCase, nCtrl = param
    sample = drawCaseControlSample(pop, cases=nCase, controls=nCtrl)
    sample.save('sample_%d.pop' % pop.dvars().gen)
    return True

pop = sim.Population(size=10000, loci=1)
pop.evolve(
    initOps=[
        sim.InitSex(),
        sim.InitGenotype(freq=[0.5, 0.5]),
    ],
    matingScheme=sim.RandomMating(),
```

```
    postOps=[
        sim.MaPenetrance(loci=0, penetrance=(0.01, 0.1, 0.15), step=100),
        sim.PyOperator(func=drawSample, param=(100, 100), step=100),
    ],
    gen = 500
)
```

A.4 EVOLVING ONE OR MORE POPULATIONS

Function `Population.evolve()` evolves a population generation by generation, following a discrete-generation evolutionary model depicted in Figure A.1. This function accepts a mating scheme and several lists of operators, which are all Python objects with their own properties and member functions. Table A.8 lists the parameters that are accepted by function `Population.evolve`.

A.4.1 Mating Scheme

A mating scheme specified by parameter `matingScheme` is used to select parents from the parental population and populate an offspring population. A mating scheme is responsible for the following:

1. *Determining the Size of the Offspring Population* The offspring population has by default the same number of individuals as the parental

TABLE A.8 Acceptable Parameters of Function `Population.evolve`

Parameter	Usage
`initOps`	A list of operators that will be applied before evolution. They are usually used to initialize a population and prepare it for evolution
`preOps`	A list of operators that will be applied to the parental population at the beginning of each generation
`matingScheme`	A mating scheme that produces an offspring population from a parental population
`postOps`	A list of operators that will be applied to the offspring population at the end of each generation
`finalOps`	A list of operators that will be applied after evolution
`gen`	Generations to evolve. Default to `-1`, which will cause the evolutionary process to continue indefinitely
`dryrun`	If set to `True`, the `evolve` function will print a description of the evolutionary process and exit

population, but you can specify a fixed population size or a dynamic population size using parameter subPopSize. In the latter case, a Python function should be defined to return the population size of the offspring population at each generation.

2. *Determining How the Parents Are Chosen* A RandomMating mating scheme chooses a male and a female parent randomly, at equal probability or at a probability that is proportional to individual fitness values if an information field named fitness exists. It is the most frequently used mating scheme in this book.
3. *Determining the Number of Offspring Per Mating Event* This is controlled by parameter numOffspring, which can be a fixed number (default to 1) or a random distribution.
4. *Determining the Sex of Offspring* This can be determined randomly or may follow a fixed pattern. Parameter sexMode is used to control this behavior.
5. *Determining How Parental Genotypes and Other Information Are Passed to Offspring* Each mating scheme has a default genotype transmitter that can be overridden with a list of operators in the parameter ops of a mating scheme.

Source code A.13 demonstrates some of the features of a simuPOP mating scheme using a monogamous mating scheme. Unlike a random mating scheme in which parents can mate with several spouses, this mating scheme chooses parents without replacement, so each parent can have only one spouse. In order to ensure equal number of male and female parents, each mating event produces exactly one male and one female offspring. A ParentsTagger is used to track the parents of each offspring so that we can determine the parents of each offspring in the simulated population. This mating scheme can be used to simulate theoretical models in which all parents pass their genotypes to the offspring population or to simulate strictly controlled mating schemes such as animal breeding programs for endangered species.

SOURCE CODE A.13 Control of the Number and Sex of Offspring in a Monogamous Mating Scheme

```
>>> import simuPOP as sim
>>> pop = sim.Population(10, loci=10, infoFields=['father_idx', 'mother_idx'])
>>> pop.evolve(
...     # use a proportion instead of probability to ensure equal numbers of
...     # males and females
...     initOps=[
```

```
...         sim.InitSex(maleProp=0.5),
...         sim.InitGenotype(freq=[0.2, 0.8])
...     ],
...     matingScheme=sim.MonogamousMating(numOffspring=2,
...         # fix the number of male offspring, the rest are females.
...         sexMode=(sim.NUM_OF_MALES, 1),
...         ops=[sim.MendelianGenoTransmitter(),
...             sim.ParentsTagger()]  # track indexes of parents
...         ),
...     gen=10,
... )
10
>>> # number of male and female offspring?
>>> sim.stat(pop, numOfMales=True)
>>> # sex of offspring?
>>> [ind.sex() for ind in pop.individuals()]
[1, 2, 1, 2, 1, 2, 1, 2, 1, 2]
>>> pop.dvars().numOfMales
5
>>> # parents of the first five offspring
>>> pop.indInfo('father_idx')[:10]
(0.0, 0.0, 2.0, 2.0, 8.0, 8.0, 4.0, 4.0, 6.0, 6.0)
```

A.4.2 Conditionally Terminating an Evolutionary Process

It is not always possible to know in advance the number of generations to evolve. For example, you may want to evolve a population until a specific allele gets fixed or lost in the population. In this case, you can let the simulator run indefinitely (do not set parameter `gen` of the `evolve` function) and depend on a *terminator* to terminate the evolution of a population. The easiest method to do this is to use population variables to track the status of a population and to use a `TerminateIf` operator to terminate the evolution according to the value of an expression.

Source code A.14 demonstrates the use of such a terminator, which terminates the evolution of a population if allele 0 at locus 5 is fixed or lost. This operator uses expression `len(alleleNum[5])==1` to determine if the allele is fixed or lost because `alleleNum[5]` is a dictionary of allele numbers, and `len(alleleNum[5])==1` implies that there is only one allele left at this locus. Source code A.14 also demonstrates the application of an interesting operator `IfElse`, which applies an operator, in this case a `PyEval`, only when an expression returns `True`.

SOURCE CODE A.14 Use a Terminator to Terminate an Evolutionary Process Conditionally

```
>>> import simuPOP as sim
>>> pop = sim.Population(50, loci=[10], ploidy=1)
>>> pop.evolve(
...     initOps=sim.InitGenotype(freq=[0.5, 0.5]),
```

```
...     matingScheme=sim.RandomSelection(),
...     postOps=[
...         sim.Stat(alleleFreq=5),
...         sim.IfElse('alleleNum[5][0] == 0',
...             sim.PyEval(r"'Allele 0 is lost at generation %d\n' % gen")),
...         sim.IfElse('alleleNum[5][0] == 50',
...             sim.PyEval(r"'Allele 0 is fixed at generation %d\n' % gen")),
...         sim.TerminateIf('len(alleleNum[5]) == 1'),
...     ],
... )
Allele 0 is fixed at generation 19
20
>>> pop.dvars().gen
20
```

A.4.3 Evolving Several Populations Simultaneously

simuPOP can evolve several populations simultaneously, which allows side-by-side comparison between instances of the same evolutionary process or evolutionary processes under slightly different settings. For example, Source code A.15 evolves three replicates of the same population simultaneously, subject to different intensity of mutations. Allele frequencies of allele 0 of three replicates are printed in a tabular format. This Source code uses a `Simulator` object, which is essentially a list of populations.

This Source code starts with the creation of a population of 5000 individuals. Instead of evolving this population directly, it creates a `Simulator` object with three replicates of this population. The `evolve` function of the simulator accepts the same set of parameters as the `evolve` function a `Population` class. The only difference is that this function evolves all populations in a simulator for one generation before it moves to the next generation.

Operators are applied to all populations in a simulator unless a parameter `reps` is used to specify indices of applicable populations. In this Source code, three `SNPMutator` operators are applied to three populations using different mutation rates. In order to print the outputs in a tabular format, a `PyEval` operator outputs a generation number once (for the first replicate), while `PyEval` operator outputs allele frequency at the first locus for each population, prefixed with a `'\t'`, followed by a newline character that is outputted by a `PyOutput` operator, which is applied only to the last replicate.

The `Simulator` class provides several functions to access its populations. Function `Simulator.population(idx)` returns a reference to the `idx`-th population in a simulator. Function `Simulator.populations()` returns an iterator that iterates through all populations in a simulator. Changing the returned population references will change populations in a simulator. If you would like an independent copy of a population, you can extract a population from a simulator using function `Simulator.extract(idx)`.

SOURCE CODE A.15 Evolve Several Replicates of a Population Simultaneously

```
>>> import simuPOP as sim
>>> pop = sim.Population(5000, loci=1)
>>> # three copies of the same population
>>> simu = sim.Simulator(pop, rep=3)
>>> simu.evolve(
...     initOps=[
...         sim.InitSex(),
...         sim.InitGenotype(freq=[0.5, 0.5])
...     ],
...     preOps=[
...         sim.SNPMutator(u=0.01, reps=0),
...         sim.SNPMutator(u=0.02, reps=1),
...         sim.SNPMutator(u=0.05, reps=2),
...     ],
...     matingScheme=sim.RandomMating(),
...     postOps=[
...         sim.Stat(alleleFreq=0),
...         sim.PyEval('gen', step=20, reps=0),
...         sim.PyEval(r"'\t%.3f' % alleleFreq[0][0]", step=20),
...         sim.PyOutput('\n', step=20, reps=-1),
...     ],
...     gen=100
... )
0 0.503 0.493 0.480
20 0.411 0.316 0.177
40 0.340 0.206 0.065
60 0.277 0.112 0.024
80 0.240 0.081 0.014
(100, 100, 100)
>>> # access populations in a simulator
>>>  for pop  in simu.populations():
...     print(pop.dvars().alleleFreq[0][0])
...
0.2108
0.0542
0.0072
>>> # extract a population
>>> pop = simu.extract(1)
>>> # print out allele frequency
>>> pop.dvars().alleleFreq[0][0]
0.0542
```

A.5 A COMPLETE simuPOP SCRIPT

Although different Python modules could be used to provide command line (e.g., `getopt`, `optparse` and `argparse`) or graphical (e.g.,

`Tkinter`) user interfaces to a simuPOP script, a simuPOP utility module `simuOpt` has been designed to provide a flexible user interface specifically for simulation studies. Instead of providing a fixed user interface, this module allows users to execute the same simuPOP script in batch mode (no user interaction), interactive mode (accept user input from command line), or through a graphical parameter input dialog. Source code A.16 lists a complete simuPOP script that makes use of this module.

The first line of this script is called a *shebang* line that tells a Linux/Unix system which interpreter should be used to execute a script. Because the Python executable might reside in different paths under different operating systems (e.g., `/usr/bin` or `/usr/local/bin`), a `/usr/env python` command is usually used to locate a Python interpreter and execute it.

The string surrounded by a pair of matching triple quotes in lines 2–8 is a module doc string. It describes the main purposes of the script and is accessible via variable `__doc__`. By passing this docstring to a `simuOpt.Param` object (line 78), this string will become part of the help message.

All examples we have seen use the standard simuPOP module. This module uses 8 bits to store an allele, so each locus can have 256 possible allelic states, ranging from 0 to 255. A run-time validation mechanism is used to monitor a simulation, which will terminate the execution of a script with a detailed error message when an invalid operation is detected. Other variants of the core simuPOP modules are provided for different applications. For example, a long allele version having 2^{32} (or 2^{64} for 64 bit operating systems) possible allele states can be used to simulate certain population genetics models such as an infinite allele model; a binary allele version that uses 1 bit for each allele should be used for diallelic (SNP) markers; and an optimized module that bypasses run-time validation can be used for production scripts that have been thoroughly tested.

Lines 10–12 demonstrate how to use the `simuOpt.setOptions` function to control which variant of simuPOP to load and how to load it. This Source code uses `alleleType='binary'` and `optimized=True` to load an optimized diallelic version of simuPOP. It also uses `quiet=True` to suppress a banner message when simuPOP is imported. Finally, option `'version=1.0.4'` indicates that this script is only compatible to simuPOP 1.0.4 and later. An error message will be displayed if an earlier version of simuPOP is imported.

SOURCE CODE A.16 A Sample simuPOP Script

```
#!/usr/bin/env python
'''
This script demonstrates the decay of linkage disequilibrium under the impact
of genetic recombination. The simulation starts with populations in which
two loci are under complete linkage disequilibrium. During evolution, parental
chromosomes are recombined before they are passed to their offspring, resulting
a gradule decay of linkage disequilibrium between these two loci.
'''
import sys
import simuOpt
simuOpt.setOptions(alleleType='binary', optimized=True, quiet=True, version='1.0.4')
import simuPOP as sim

options = [
    {'name': 'popSize',
     'default': 1000,
     'type': int,
     'label': 'Population Size',
     'validator': 'popSize > 0',
    },
    {'name': 'gen',
     'default': 50,
     'type': int,
     'label': 'Generations to evolve',
     'validator': 'gen > 0',
    },
    {'name': 'recRate',
     'default': 0.01,
     'type': (int, float),
     'label': 'Recombination Rate',
     'validator': 'recRate >= 0. and recRate <= 0.5',
    },
    {'name': 'numRep',
     'default': 5,
     'type': int,
     'label': 'Number of Replicate',
     'validator': 'numRep > 0',
    },
    {'name': 'measure',
     'default': "D'",
     'type': ('chooseOneOf', ["D'", 'R2']),
     'label': 'LD measure',
     'description': '''A measure of linkage disequilibrium to be outputed.
        Acceptable input includes
        |D': Lewontin's D' measure which is the standard LD measure divided
        by the theoretical maximum for the observed allele frequencies.
        |R2: R2 is the correlation coefficient between pairs of loci.
        ''',
    },
]

def simuLDDecay(popSize, gen, recRate, numRep, measure):
    '''Simulate the decay of linkage disequilibrium as a result
    of recombination.
    '''
    simu = sim.Simulator(
        sim.Population(size=popSize, ploidy=2, loci=[2]),
        rep=numRep)
    simu.evolve(
        initOps=[
            sim.InitSex(),
            sim.InitGenotype(haplotypes=[[0, 1], [1, 0]])
        ],
```

```
        matingScheme=sim.RandomMating(ops=sim.Recombinator(rates=recRate)),
        postOps=[
            sim.Stat(LD=[0, 1]),
            sim.IfElse(measure=="D'",
                sim.PyEval(r"'%.4f\t' % LD_prime[0][1]"), # if measure=="D'"
                sim.PyEval(r"'%.4f\t' % R2[0][1]")),      # if measure=="R2"
            sim.PyOutput('\n', reps=-1),
        ],
        gen = gen
    )

if __name__ == '__main__':
    # get all parameters
    pars = simuOpt.Params(options, '''A demonstration of the decay of linkage
        disequilibrium with the impact of genetic recombination.''', __doc__)
    if not pars.getParam():
        sys.exit(0) # cancelled or -h, --help
    # call the simulation function
    simuLDDecay(pars.popSize, pars.gen, pars.recRate, pars.numRep, pars.measure)
```

Lines 14–50 define five parameters `size`, `gen`, `recRate`, `numRep`, and `measure` using a list of parameter specification dictionaries. These dictionaries can have keys `name`, `default`, `type`, `label`, `description`, and `validator`, which are used to obtain, validate, and convert user inputs for each parameter (Table A.9). For example, because the type of parameter `numRep` is `int`, simuPOP will try to convert user inputs to an integer (e.g., from string `'50'` or `'5*10'` to `50`) and reject invalid inputs such as `50.5`. Similarly, parameter `measure` will only accept one of the two specified values `D'` and `R2`. In addition to one or more allowed Python types, the `type` field also accepts values such as `'numbers'`,

TABLE A.9 Acceptable Keys in a Parameter Specification Dictionary

Field	Usage
`name`	Name of the parameter
`default`	Default value for this parameter
`type`	Type of acceptable input, which help simuPOP determine the GUI widget for a parameter, how to convert user input to appropriate format, and how to validate a parameter
`label`	A label to display in the parameter input dialog. If this field is missing, a parameter will not be displayed in the parameter input dialog
`description`	A detailed description of the parameter, which will be displayed as tooltip of the parameter in the parameter input dialog and be used to generate help messages of the script
`validator`	A function or an expression to validate a user input

and 'filename', which accept a list of integer or float numbers, and a valid filename, respectively.

Although the type of parameter provides type validation for user inputs, a validator can be used to provide further validations. This item accepts a function or a Python expression. When a function is provided, it will be applied to a user input. A user input will be rejected if this function returns False. For example, if a parameter defines a probability, a function returned by function simuOpt.valueBetween(0,1) will return True only if the input is between 0 and 1.

This example uses expressions to validate parameters. These expressions are evaluated in a dictionary where variables are defined from names and values of parameters. A parameter will be rejected if its validating expression returns False. For example, expression 'gen > 0' will reject 0 as an input for parameter gen. This method is very flexible in that information from other parameters can be used to validate a parameter. For example, if two parameters opt1 and op2 are required to have the same length, expression 'len(opt1) == len(opt2)' can be used to validate both parameters.

Lines 52–73 define a function simuLDDecay, which is the main simulation function of this script. This function is an extension to Source code A.1. It allows the simulation of several replicates of the population simultaneously and also allows the display of two linkage disequilibrium measures.

Function simuLDDecay will only be executed if the script is executed as a script (if __name__ == '__main__', line 75), not imported as a module. This is a very useful feature of Python because the same script can be executed directly or be used as a Python module to provide functions to another module. For example, if a user would like to call function simuLDDecay with different parameters, he or she can import this script as a module and call this function directly as follows:

```
from simuLDDecay import simuLDDecay
for r in [0.1, 0.01, 0.002]:
    simuLDDecay(1000, 50, r, 2, 'R2')
```

The last block (lines 75–82) is the execution part of this script. It defines a simuOpt.Params object using the parameter specification list options, a short description, and a detailed description __doc__. A function pars.getParam() is used to obtain values of parameters size, gen, recRate, numRep, and measure. If successful, parameter values are passed to function simuLDDecay to perform the simulation.

There are a number of advantages of using the `simuOpt.Params` class to handle user inputs. By specifying the type of parameters, this class automatically converts user inputs into required types. Arbitrary Python expressions are allowed so that users can use expressions such as `range(10)` to input long arguments. If a list of values (e.g., `numbers`) is needed, single input will be converted to a list automatically. The `simuOpt.Params` also validates user input and will reject a value if it is not of required type or does not pass specified validation function or expression.

The `getParams()` function of the `Params` class does all the hard job of obtaining values of parameters from command line, configuration file, a graphical parameter input dialog, or interactive user input. It first checks command line for argument `-h` or `-help` and will print a help message and return `False` if one of them is specified. This usage message is created from descriptions of the script and parameters. The help message of script simuLDDecay.py is listed below.

Before falling into a particular mode to collect user input, the `getParams` function processes command line arguments such as `-gen=50` to obtain values for parameters. Parameters of a Boolean type can be set to `True` by parameter `-param`. If a configuration file is specified by command line argument `-config`, this function will try to obtain values of parameters from this file. A configuration file can be created manually, but is most frequently saved by function `Params.saveConfig(filename)`.

The `getParams` function then checks for command line option `-gui` to determine which user interface to use. If the script is executed under a batch mode (`-gui=batch`), this function will return `True` directly, so default values will be used for parameters that are not specified from command line or a configuration file. For example, command

```
> simuLDDecay.py --gui=batch --size=5000
```

will execute the script with specified population size and default parameters of all other parameters. In contrast, if the script is executed under an interactive mode (`-gui=interactive`), if running in an interactive mode, this function will prompt users for values of parameters that have not been specified through command line or a configuration file.

If no gui mode is specified, the script will be executed in GUI mode where a parameter input dialog will be displayed for users to view and edit parameters (Figure A.2). Different GUI widgets such as checkbox and listbox will be used for different types of parameters. Detailed descriptions of parameters are displayed as tooltips to these parameters. A parameter

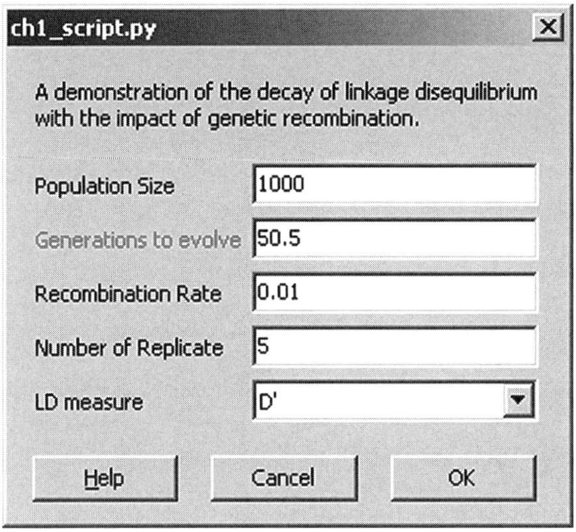

FIGURE A.2 Parameter input dialog for Source code A.16. A parameter input dialog with five parameters. The label of the second parameter is marked in red when the OK button is clicked because its input value is not of required type (`int`).

will be marked in red if an invalid parameter is detected when the OK button is pressed.

Values of each parameter can be accessed as attributes of the `Param` object if function `pars.getParam()` returns `True`. Line 82 of Source code A.16 uses this feature to pass values to function `simuLDDecay` to execute the main simulation program. Alternatively, values of all parameters can be returned as a list using function `Param.asList` or as a dictionary using function `Param.asDict`. Because parameter order and names of the `Param` object match those of the `simuLDDecay` function, we can also pass parameters to this function using

```
simuLDDecay(*pars.asList())
```

or

```
simLDDecay(**pars.asDict())
```

We prefer the method used in Source code A.16 because it is less error prone.

In summary, despite the availability of other parameter-handling modules, we recommend the use of module `simuOpt` because it allows a self-documentary method to describe parameters and a flexible mechanism to execute the same script in both GUI and batch mode. However, for the sake of brevity, we do not use this style in examples we describe in this book.

REFERENCES

1. International HapMap Consortium, A haplotype map of the human genome. *Nature*, 437(7063):1299–1320, 2005.

INDEX

Forward-time Population Genetics Simulations: Methods, Implementation, and Applications,
Bo Peng, Marek Kimmel, and Christopher I. Amos.